Farewell to Reason

Farewell to Reason

PAUL FEYERABEND

VERSO

London · New York

First published by Verso 1987
© Paul Feyerabend
All rights reserved

Verso
UK: 6 Meard Street, London W1V 3HR
USA: 29 West 35th Street, New York, NY 10001 2291

Verso is the imprint of New Left Books

British Library Cataloguing in Publication Data

Feyerabend, Paul
  Farewell to reason.
  1. Relativity   2. Rationalism
  I. Title
153.4'3   BD221

ISBN 0-86091-184-5
ISBN 0-86091-896-3 Pbk

US Library of Congress Cataloging in Publication Data

Feyerabend, Paul K., 1924-
    Farewell to reason/Paul Feyerabend.
    p.   cm.
    Collection of essays rewritten for publication in book form.
    ISBN 0-86091-184-5: $39.95 (U.S.). ISBN 0-86091-896-3 (pbk.): $14.95
(U.S.)
    1. Philosophy.   I. Title.
  B29.F448 1987
  100—dc19

Printed in Great Britain by Biddles Ltd, Guildford

Contents

Introduction

The essays collected in this volume deal with cultural diversity and cultural change. They try to show that diversity is beneficial while uniformity reduces our joys and our (intellectual, emotional, material) resources.

There exist powerful traditions which oppose such a point of view. They concede that people may arrange their lives in various ways; however they add that there must be limits to variety. The limits, they say, are constituted by moral laws which regulate human actions and by physical laws which define our position in nature. Philosophers from Plato to Sartre and scientists from Pythagoras to Monod have claimed to possess such laws and have complained about the variety (of values, beliefs, theories) that still remained.

In the late seventies and the early eighties these complaints increased in urgency. 'Contemporary culture', we hear, is in a 'crisis'. It is torn apart by a 'profound contradiction between the traditional humanist views of man and the world on the one hand, and the value devoid mechanistic description of science on the other' while the humanities, philosophy, the arts, and social thought are in addition corroded by a 'cultural cacophony', or a 'philosophical malady'. So extreme is the fragmentation perceived by some critics that Jurgen Habermas recently spoke of a 'neue Unübersichtlichkeit' – a new unsurveyability: it is impossible to find one's way in the flood of styles, theories, points of view, that inundate public life. [1]

1. For the 'crisis' and the 'profound contradiction' see Roger Sperry, *Science and Moral Priority*, New York 1985, p. 6. For the clash between the sciences and

1

The complaints are surprisingly uninformed. It is true that 'culture' occasionally gets somewhat disorganised. But the trend is not new and it is balanced by powerful contrary tendencies: schools outflank, overrun, absorb other schools, scientists from different fields create interdisciplinary subjects (examples: synergetics; molecular biology), grand 'unifying schemes (evolution; holism; dualistic solutions of the mind-body problem; linguistic speculations) blur important distinctions, film, computer art, rock music, high energy physics (cf. Andrew Pickering's fascinating essay *Constructing Quarks*, Chicago 1985) combine business principles, artistic inspiration and scientific discovery in a manner reminiscent of 15th century Renaissance practice. There is fragmentation; but there are also new and powerful uniformities.

The lack of perspective displayed by the critics is even more astonishing. They speak of 'the (!) crisis of contemporary culture' or of 'world culture'; what they mean is Western academic and artistic life. But the quarrels of professors and the contortions of Western art shrink into insignificance when compared with the steady expansion of Western 'progress' and 'development' which is the spreading of Western business, science and technology. This is an international phenomenon; it characterises capitalist as well as socialist societies; it is independent of ideological, racial and political differences and it affects an increasing number of peoples and cultures. There is hardly any trace here of the debates and disagreements that so exercise our intellectuals. What is being imposed, exported and again imposed is a collection of uniform views and practices which have the intellectual and political support of powerful groups and institutions. [2] By now Western forms of life are found in the most remote corners of the world and have changed the habits of people who only a few decades ago were unaware of their existence. Cultural differences disappear, indigenous crafts, customs, institutions are being replaced by Western

humanities cf. C.P. Snow, *The Two Cultures and the Scientific Revolution*, Cambridge 1959 and W.T. Jones, *The Sciences and the Humanities*, Univ. of California Press 1965. Jones calls the rift between the sciences and the humanities 'the crisis of contemporary culture'. The phrase 'cultural cacophony' occurs in an issue of the journal *Precis*, introducing a discussion of postmodernism at the Columbia University School of Architecture; the 'malady' in K.R. Popper, *The Open Society and its Enemies*, Vol. 2, New York 1966, p. 369. Habermas's observations are found in *Die Neue Unübersichtlichkeit*, Frankfurt 1985.

2. Decisive among these groups is a new class, a scientific-technological elite which, according to some writers (Daniel Bell and John Kenneth Galbraith

objects, customs, organizational forms. The following passage from a Presidential Address to the American Society of Parasitology contains an excellent description of this process:

> The essence of preindustrial indigenous societies is their variety and local adaptation. Each is tied to a specific habitat and has evolved its own cultural and behavioral expression. The wide variety of resulting human social forms is a response to an equal variety of habitats, each with a set of distinctive environmental constraints.
>
> In almost diametric opposition, industrial technological development is characterised by a controlled, relatively uniform and highly simplified environment, typically with a widespread reduction in the number of species to a few domesticated forms, including humans, to accidentally domiciliated plants and animal weeds . . . High levels of environmental pauperization and widely distributed homogenization characterise industrialised societies in all political and economic systems of the world (Donald Heyneman, *Journal of Parasitology*, 70(1), 1984, p.6).

'We are . . . threatened by monotony and dullness' is François Jacob's succinct summary (*The Possible and the Actual*, Seattle and London 1982, p. 67). Western civilization itself has been losing its diversity to such an extent that an American author could write, in the April 18, 1985 issue of the *International Herald Tribune*: 'Like an advancing fog sameness is engulfing the country' (the USA). The conflicts which so displease our cultural critics become invisible when compared with this massive trend towards natural, social and technological uniformity.

The trend is not beneficial, even when judged by the values of those who have hitherto encouraged it. There are ecological

---

among them), increasingly determines prestige and power. Bakunin, who emphasized the importance of scientific knowledge, also warned against 'the reign of scientific intelligence, the most aristocratic, despotic, arrogant and elitist of all regimes' (*Bakunin on Anarchy*, Sam Dolgoff tr., New York 1972, p. 319). Today his fears have become reality. Even worse, knowledge has become a commodity, its legitimacy linked to the legitimation of the legislator: 'science seems more completely subordinated to the prevailing powers than ever before and . . . is in danger of becoming a major stake in their conflict', J.F. Lyotard, *The Postmodern Condition, A Report on Knowledge*, Minneapolis 1984, p.8. The elite administering the community often supports what E.P. Thompson called 'exterminism', a structure of abstract research and technological development directed towards mass killings: E.P. Thompson et al. eds., *Exterminism and Cold War*, London 1982, pp. 1ff, especially p. 20. Cf. also N. Chomsky, 'Intellectuals and the State', in *Towards a New Cold War*, New York 1986, as well as the debates about the function of the National Laboratories reprinted in the *Bulletin of the Atomic Scientists* of 1985.

problems. They are world wide, well documented, fairly well understood and many people know them from personal experience (chemical and radioactive pollution of rivers, oceans, the air, the ground water; depletion of the ozone layer; a drastic decrease in the number of animal and plant species; land desertification and deforestation). Many so-called 'Third World problems' such as hunger, illness and poverty seem to have been caused rather than alleviated by the steady advance of Western Civilization.[3] The spiritual impact of the trend is less obvious but not less painful. For many societies acquiring knowledge was part of living; the knowledge acquired was relevant and reflected personal and group concerns. The imposition of schools, literacy and 'objective' information detached from local preferences and problems emptied existence of its epistemic ingredients and made it barren and meaningless. Here, too, the West led the way by separating schools from life and subjecting life to scholastic rules.[4]

Studying phenomena such as these, scholars, representatives of indigenous cultures, and international associations concluded that there are many ways of living, that cultures different from our own are not mistakes but results of a delicate adaptation to particular surroundings, and that they found, rather than missed, the secrets of a good life. Even highly technical problems such as the problems of arms control are never entirely 'objective' but are pervaded by 'subjective', i.e. cultural components.[5] 'In humans', writes François Jacob in a passage from which I already quoted,

natural diversity is . . . strengthened by cultural diversity, which

3. A general account is John H. Bodley, *Victims of Progress*, Menlo Park, California 1982. The special cases of health and hunger are analysed by Grazia Borrini, 'Health and Development – A Marriage of Heaven and Hell?' in A. Ugaldo, ed., *Studies in Third World Society*, Austin, Texas 1986. M. Rahnema, 'From "Aid" to "Aids" – a Look at the Other Side of Development', manuscript, Stanford 1983 describes how the imposition of Western technology destroyed social immune systems that had provided efficient protection against natural and social catastrophes. F.A. von Hayek's warning that societies which had grown through a long process of adaptation were better equipped to deal with problems than intellectuals using the most advanced theories and equipment at their disposal was drastically confirmed by computer models of the effects of 'rational interventions' – they all left the society in a worse state than beffore: F.A. von Hayek, *Missbrauch und Verfall der Vernunft*, Salzburg 1979.'

4. Cf. M. Rahnema, 'Education for Exclusion or Participation?', ms, Stanford, April 16, 1985. The situation in the West is described in I. Illich, *Deschooling Society*, New York 1970.

allows mankind to better adapt to a variety of life conditions and to better use the resources of the world. In this area, however, we are now threatened with monotony and dullness. The extraordinary variety which humans have put into their beliefs, their customs and their institutions is dwindling every day. Whether people die out physically or become transformed under the influence of the model provided by industrial civilization, many cultures are disappearing. If we do not want to live in a world covered with a single technological, pidgin-speaking, uniform way of life – that is, in a very boring world – we have to be careful. We have to use our imagination better (Jacob, p. 67).

The essays assembled in the present volume support this point of view by criticizing philosophies that oppose it.

More especially, I shall criticize two ideas that have often been used to make Western expansion intellectually respectable – the idea of Reason and the idea of Objectivity.

To say that a procedure or a point of view is objective(ly true) is to claim that it is valid irrespective of human expectations, ideas, attitudes, wishes. This is one of the fundamental claims which today's scientists and intellectuals make about their work. *The idea of objectivity*, however, is older than science and independent of it. It arose whenever a nation or a tribe or a civilization identified its ways of life with the laws of the (physical and moral) universe and it became apparent when different cultures with different objective views confronted each other. There are various reactions to such an event; I mention three.

One reaction was *persistence*: our ways are right and we are not going to change them. Peaceful cultures tried to avoid change by avoiding contact. The pygmies, for example, or the Mindoro of the Philippines did not fight Western intruders, they did not

5. In his book *Facing the Threat of Nuclear Weapons*, Seattle and London 1983, pp. 36ff Sidney Drell, advisor to the US government on issues of national security and arms control, states four 'negotiating goals'. The third goal is 'To allow the two countries [the USA and the USSR] to implement the negotiated provisions by means of selective reductions in accord with their very different technological and bureaucratic styles. While these reductions must be equitable, they may, at the same time, be assymetric. Thus the negotiations must be highly flexible'. Rudolf Peierls, *Bird of Passage*, Princeton University Press 1985, p. 287 writes in a similar vein: 'It is evident that the geographical and strategic situations of the two sides are different; that their nuclear weapons, the means of delivering them, and their intelligence organizations are completely different; and that therefore any evaluation of the relative strength becomes highly speculative': an exchange between two nations that touches their life interests cannot be carried out in an 'objective' and schematic way.

submit to them either, they simply moved out of their sphere of influence. More belligerent nations used war and murder to eradicate what did not fit their vision of the Good. 'The Law of Moses', writes Eric Voegelin on this point, 'abounds with blood-thirsty fantasies concerning the radical extermination of the goyim in Canaan at large, and of the inhabitants of cities in particular. And the law to exterminate the goyim is . . . moti-vated by the abomination of their adherence to other gods than Yahweh: the wars of Israel in Deuteronomy are religious wars. The conception of war as an instrument for exterminating every-body in sight who does not believe in Yahweh is an innovation of Deuteronomy. . . .' (*Order and History* Vol. i, *Israel and Reve-lation*, Louisiana State University Press 1956, pp. 375f). The representatives of Western civilization, though fond of humani-tarian slogans, were not always averse to such conceptions.

Persistence also characterizes more recent developments in the (physical and social) sciences which are holistic, emphasize historical processes instead of universal laws and let 'reality' arise from an (often indivisible) interaction between observer and the things observed. For the authors who encourage the trend (Bohm, Jantsch, Maturana, Prigogine, Varela, the proponents of an 'evolutionary epistemology' and others) defuse cultural variety by showing that and how it fits into their scheme. Instead of providing guidance for personal and social choices they with-draw into their theoretical edifices and explain from there why things were as they were, are as they are and will be as they will be. This is the old objectivism all over again, only wrapped in revolutionary and pseudo-humanitarian language.

A second reaction is *opportunism*: the (leaders of the) con-flicting cultures examine each others' institutions, customs, beliefs and accept or adapt those they find attractive. This is a complex process. It depends on the historical situation, the atti-tudes of the participants, their fears, needs, expectations; it may lead to power struggles inside the cultures, it may even be influ-enced by temporary changes of the weather which magnify some faults and diminish others. At any rate: an opportunistic en-counter of cultures cannot be reduced to general rules. Cultural opportunism was practiced (and is still being practiced) by in-dividuals, small groups and entire civilizations. An example is the case of the nations, kingdoms, tribes that populated the Ancient Near East during the late Bronze Age, a period the Egyptologist Henry Breasted called the 'First Internationalism'. These nations, kingdoms, and tribes were often at war with each other but they exchanged materials, languages, industries, styles, people with special skills such as architects, navigators, prostitutes – and even gods (details are given in T.B.L. Webster, *From Mycenae to Homer*, New York 1964). Another example is the Mongolian Empire. Marco Polo testifies to the enormous curiosity the successors of Ghengis Khan showed for foreign things and to the ingenious ways in which they adapted them to their needs. The Great Khan himself realised the importance of foreign customs and opposed advisers who wanted to destroy cities in order to universalise their own nomadic habits. An interesting modern example are the Natives of Karen Blixen's Kenya (see Isak Dinesen, *Out of Africa,* New York 1985, pp. 54f.) who 'due . . . to their acquaintance with a variety of races and tribes' were 'more [men] of the world than the suburban or provincial settler[s] who ha[d] grown up in a uniform community and with a set of stable ideas.'

A third reaction is *relativism*: customs, beliefs, cosmologies are not simply holy, or right, or true; they are useful, valid, true *for* some societies, useless, even dangerous, not valid, untrue *for* others. As I shall explain in chapter 1, relativism has many forms, some more intuitive and unreflected, others highly intellectual. It is more widespread than the critics of the intellectual versions seem to think. For example, the pygmies who tried to escape Western ways may well have been relativists and not dogmatists – assuming they cared about distinctions such as these (see C.M. Turnbull, 'The Lesson of the Pygmies', *Scientific American* 208(1), 1963).

According to many historians, the Greeks introduced still another method for dealing with cultural variety. Trying to separate the Right Way from the Wrong Way, they relied neither on firmly held traditions, nor on ad-hoc adaptations; they relied on *argument*.

Now argument was not a new invention. Argument occurs in all periods of history and in all societies. It plays an important role in the opportunistic approach: an opportunist must ask himself how foreign things are going to improve his life and what other changes they will cause. Occasionally 'primitives' used argument to turn the tables on anthropologists who tried to convert them to rationalism. 'Let the reader consider any argument that would utterly demolish all Zande claims for the power of [their] oracle[s]', writes E.E. Evans-Pritchard on this matter (*Witchcraft, Oracles and Magic Among the Azande*, Oxford 1937, pp. 319f). 'If it were translated into Zande modes of thought it would serve to support their entire structure of belief. For their mystical notions are eminently coherent, being inter-related by a network of logical ties and are so ordered that they never too crudely contradict sensory experience but, instead, experience seems to justify them.' 'I may remark', Evans-Pritchard adds (p. 270), 'that I found this [i.e. consulting oracles for day-to-day decisions] as satisfactory a way of running my home and affairs as any other I know of.' Argument, like language, or art, or ritual, is universal; but, again like language or art or ritual, it has many forms. A simple gesture, or a grunt can decide a debate to the satisfaction of some participants while others need long and colourful arias to be convinced. Thus argument was well established long before the Greek philosophers started thinking about the matter. What the Greeks did invent was not just argument, but a special and standardised way of arguing which, they believed, was independent of the situation in which it occurred and whose results had universal authority. Thus the ancient idea of tradition-independent *truths* (the material notion of objectivity, as it could be called) which had run into the problem of cultural variety was replaced by the somewhat less ancient idea of tradition-independent *ways of finding truths* (the formal notion of objectivity). And being rational or using reason now meant using these ways and accepting their results.

The formal notion of objectivity has problems similar to those of the material notion. This is not surprising – 'formal' procedures make sense in some worlds, they become silly in others. For example, the demand for unending criticisms followed by increasingly comprehensive explanations breaks down in a universe that is finite, both qualitatively and quantitatively. Being liable to disrupt forms of life that provide material security and spiritual fulfillment it may be rejected for ethical reasons as well: some people prefer flourishing in a stable world to constantly adapting to new ideas. The demand to look for refutations and to take them seriously leads to an orderly development only in a world where refuting instances are rare and turn up at large intervals, like large earthquakes. In such a world we can improve, build, live peacefully with our theories from one refutation to the next. But all this is impossible if theories are surrounded by an 'ocean of anomalies' as is the case in most social matters (cf. my *Philosophical Papers*, Cambridge, Mass. Vol. 1, chapter 6, section 1). The habit of objectivizing basic beliefs or the results of research becomes nonsensical in a world that contains some form of complementarity and in societies adapted to close social contacts. And non-contradiction cannot be demanded in a world where an old woman is seen as having 'the round, sweet throat of a goddess' (*Iliad* 3.386ff). The result is clear: cultural variety cannot be tamed by a formal notion of objective truth because it contains a variety of such notions. Those who insist on a particular formal notion are just as liable to run into problems (in their sense) as the defenders of a particular conception of the world.

As science advanced and produced a steadily increasing store of information, formal notions of objectivity were used not only to *create knowledge,* but also to *legitimize,* i.e. to show the objective validity of, *already existing bodies of information.* This led to further problems: there exists no finite set of general rules that has substance (i.e. recommends or forbids some well-defined procedures) *and* is compatible with all the events leading to the rise and progress of modern science. Formal requirements defended by scientists and philosophers were found to be in conflict with developments set in motion and supported by the same group. To resolve the conflict the requirements were gradually weakened until they disappeared into thin air.[6]

Scientists also undermined universal principles of research in a more direct way. Who would have thought that the boundary between subject and object would be questioned as part of a scientific argument and that science would be advanced thereby? Yet this was, precisely what happened in the quantum theory, in physiological studies such as those of Maturana and Varela (and earlier, in Mach's investigations on the physiology of perception). Who would have thought that the 'repugnant' (Eddington) and 'theological' (Hoyle) notion of a beginning of the universe would again play an important role? Yet Friedmann's calculations and the discoveries of Hubble and others had precisely this result. Who would have thought that scientific theories could be upheld in the face of unambiguous negative evidence and that science would profit from such a procedure? But Einstein, who more than once ridiculed the concern for a 'verification of little

6. The weakening process is described with flair and historical examples in I. Lakatos, 'Falsification and the Methodology of Research Programmes', in I. Lakatos and A. Musgrave, eds., *Criticism and the Growth of Knowledge*, Cambridge 1970; cf. the same author's 'History of Science and Its Rational Reconstructions', *Boston Studies in the Philosophy of Science*, vol. viii. For the endpoint, the disappearance of all content, cf. Vol. ii, chapter 8, section 9 and chapter 10 of my *Philosophical Papers*.

effects', made progress in precisely this manner. And a look at research periods such as the older quantum theory or at the developments that preceded the discovery of the structure of DNA shows that the idea of a science that proceeds by logically rigorous argumentation is nothing but a dream.[7] Of course, there is rigour in all these prima facie chaotic procedures just as there is rigour in the *Demoiselles d'Avignon* – but it is a rigour that fits the situation, is complex, changes and differs greatly from the 'objective' rigour of our less gifted logicians and epistemologists.[8]

A second idea that plays an important role in the defence of Western civilization is *the idea of Reason* (with a capital 'R') or rationality. Like the notion of objectivity, this idea has a material and a formal variant. To be rational in the material sense means to avoid certain views and to accept others. For some intellectuals among the early Christians, Gnosticism, with its colourful hierarchies and its strange developments, was the height of irrationality. Today being irrational means, for example, believing in astrology, or creationism, or, for different groups, believing in the racial origin of intelligence. To be rational in the formal sense again means to follow a certain procedure. Hardnosed empiricists regard it as irrational to retain views plainly in conflict with experiment while hardnosed theoreticians smile at the irrationality of those who revise basic principles at every flicker of the evidence. These examples already show that it would be hardly fruitful to let statements such as 'this is rational' or 'this is irrational' influence research. The notions are ambiguous and never clearly explained, and trying to enforce them would be counter productive: 'irrational' procedures often lead to success (in the sense of those who call them 'irrational'), while 'rational' procedures may cause tremendous problems. Strictly speaking we have here two *words*, 'Reason' and 'Rationality', which can be connected with almost any idea or procedure and then surround it with a halo of excellence. But how did the two words manage to get their enormous beatifying power?

The assumption that there exist universally valid and binding standards of knowledge and action is a special case of a belief whose influence extends far beyond the domain of intellectual

7. This applies to the 'context of discovery' *and* the 'context of justification': good justifications have to be discovered just like good theories or good experiments.

8. For details cf. chapter 5, Vol. ii of my *Philosophical Papers*, Cambridge, Mass. 1981.

debate. This belief (of which I have already given some examples) may be formulated by saying that there exists a right way of living and that the world must be made to accept it. The belief propelled the Moslem conquests; it accompanied the crusaders into their bloody battles; it guided the discoverers of new continents; it lubricated the guillotine and it now provides fuel for the endless debates of libertarian and/or Marxist defenders of Science, Freedom and Dignity. Of course, each movement filled the belief with its own particular content; it changed the content when difficulties arose and it perverted it when personal or group advantages were at stake. But the idea that there is such a content, that it is universally valid and that it justifies intervention always played and is still playing an important role (as I indicated above, this is held even by some critics of objectivism and reductionism.) We may surmise that the idea is a leftover from times when important matters were run from a single centre, a king or a jealous god, supporting and giving authority to a single world view. And we may further surmise that Reason and Rationality are powers of a similar kind and are surrounded by the same aura as were gods, kings, tyrants and their merciless laws. The content has evaporated; the aura remains and makes the powers survive.

The absence of content is a tremendous advantage; it enables special groups to call themselves 'rationalists', to claim that widely recognized successes were the work of Reason and to use the strength thus gained to suppress developments contrary to their interests. Needless to say, most of these claims are spurious.

I have already mentioned the case of the sciences: they may proceed in an orderly way but the patterns that occur are not stable and cannot be universalised. Enlightenment, another alleged gift of Reason, is a slogan, not a reality. 'Enlightenment', wrote Kant ('What is Enlightenment?', cited in L.W. Beck, ed, *Kant, On History*, Library of Liberal Arts 1957, p. 3), 'is man's release from his self-incurred immaturity. Immaturity is man's inability to make use of his understanding without direction from another. Self-incurred is this immaturity when its cause lies not in lack of reason but in lack of resolution'. Enlightenment in this sense is a rarity today. Citizens take their cue from experts, not from independent thought. This is what is now meant by 'being rational'. Increasing parts of the lives of individuals, families, villages, cities are taken over by specialists. Very soon a person will not be able to say 'I am depressed' without having to listen to the objection, 'So you think you are a psychologist?'.[9] 'If I have a book', Kant wrote long ago, 'which understands for me, a pastor who has a conscience for me, a physician who decides my diet and so forth, I need not trouble myself. I need not think, if I can only pay – others will readily undertake the irksome work for me.' It is true that there is, and always was reason (with a small r) for hope. There always exist people who fight uniformity and defend the right of individuals to live, think, act as they see fit. Entire societies, 'primitive' tribes among them, have taught us that the progress of Reason is not inevitable, that it can be retarded and that things may get better as a result. Scientists are moving beyond traditional boundaries of research, citizens both in small communities and on a large scale examine expert decisions that affect them. But these are small favours compared with the increasing centralization of power that is an almost unavoidable consequence of large-scale technologies and corresponding institutions.

Which brings me to the matter of *liberty*. The comfortable liberties we have and which millions are still lacking have only rarely been achieved in a Reasonable way – they came from fights and compromises that took many elements into account and whose structure cannot be explained by general principles, covering all cases, from Cleisthenes to Mandela. Each movement had its own policy, its own feelings, its own imagination; each struggle gave its own meaning to this one powerful word: freedom. It is of course possible to summarise the individual struggles and to extract a 'political lesson' from the summaries – but without concrete policies the result is always bland, useless, misleading, unrealistic. It is also possible to use the most vapid slogans and the most empty 'principles' to sell or impose a coherent and meaningful view of the world. This does not encourage freedom, it breeds slavery, though slavery packaged in resounding libertarian phrases.

Combining these considerations with the insights gained by scientists studying the material and spiritual achievements of indigenous peoples, we find that *there is nothing in the nature of science that excludes cultural variety*. Cultural variety does not conflict with science viewed as a free and unrestricted inquiry, it

9. In a surprising passage (*Theaet.* 144d8–145a13) which anticipates the whole development, Plato's Socrates criticizes Theodoros, one of the interlocutors of the dialogue, for having said that Theaetetus looked like him although he is no expert in the recognition of facial similarities.

conflicts with philosophies such as 'rationalism' or 'scientific humanism' and an agency, sometimes called Reason, that use a frozen and distorted image of science to get acceptance for their own antediluvian beliefs. But rationalism has no identifiable content and reason no recognizable agenda over and above the principles of the party that happens to have appropriated its name. All it does now is to lend class to the general drive towards monotony. It is time to disengage Reason from this drive and, as it has been thoroughly compromised by the association, to bid it farewell.

What I have told so far is one side of the story: many things were achieved despite Reason, not with its help. The other side is that Reason did leave its mark. It distorted the achievements, stretched them beyond their limits, and is therefore at least in part responsible for the excesses that are being propagated under its name. My arguments in the following essays will deal with the false consciousness created by the presence of this distorting agency.

I start with a philosophy that undermines the very basis of Reason, namely relativism. For the purpose of discussion I have dissolved the monolith 'relativism' into a series of theses, starting with modest and almost trivial assertions (though there exist objections even here!) and proceeding to more daring and, unfortunately, also more technical assertions. My aim is to show that relativism is reasonable, humane and more widespread than is commonly assumed.

Next, in chapter 2, I present Xenophanes, the First Western Intellectual. Xenophanes was an interesting character, a lively intellect; he was not averse to jokes, nor to pompous statements and heavy rhetoric. His criticism of tradition and especially his criticism of the Homeric gods has received praise from writers as different as Mircea Eliade, W.K.C. Guthrie, Karl Popper and Franz Schachermayr. It shows at once the basic dishonesty of all Rational philosophies: they introduce strange assumptions which are neither plausible nor argued for, and then ridicule opponents for holding different views. Xenophanes' immediate successors noticed this weakness. Herodotus and Sophocles wrote about the gods as if Xenophanes had never existed and some early scientists criticized the abstract approach in their own field.

This brings me to the subject of chapter 3, the idea that knowledge should be based on universal principles or theories. It cannot be denied that there *are* successful theories using rather abstract concepts. But before making their existence the basis of far-reaching conclusions about everything we claim to know, we must ask the following three questions: what do these theories mean (do they describe pervading features of an 'objective reality' or do they merely aid us in the prediction of events whose nature is determined independently?); how effective are they (perhaps theories, taken literally, are always inadequate and the true consequences ascribed to them are produced with the help of corrective ad hoc assumptions, so-called approximations?); and – how are they used? Having answered these questions, we must further ask if the successes of physics, astronomy or molecular biology imply that medicine, national defence or our relations with other cultures should include principles of comparable objectivity and abstractness.

The answer to the third question – how are scientific theories used? – is that the practice of inventing, applying and improving theories is an art and therefore a historical process. Science as a living enterprise (as opposed to science as a 'body of knowledge') is part of history. The formulae that adorn our textbooks are temporarily frozen parts of activities that move with the stream of history. They must be melted down, reconnected with the stream, in order to be understood and to produce results. They *are* melted down whenever there is a fundamental change, as is shown by the scientific papers of such periods. Distinctions such as the distinction between the natural sciences and the social sciences, or the older distinction between Naturwissenschaften and Geisteswissenschaften, or the related distinction between the sciences and the arts (the sciences and the humanities), are not distinctions between real things, but between real things (arts, humanities, sciences – all of which are, or deal with, traditions-in-use) and nightmares about them.

The nightmares had a decisive and by no means benevolent influence on our attempts to understand knowledge. If reality is described by theories that are not only valid independently of human life but contain none of its features, then how is it possible for the human mind ever to reach it? Would there not be an unbridgeable gulf between human efforts and their alleged results? But the human mind does reach reality – the success of the sciences testifies to that. Attempts to provide a Rational account of the bridging process (theories of induction and confirmation; transcendental idealism) were unsuccessful. They also seemed to turn scientists into inductive machines and data-processors. Then the champions of theory suggested a simple solution: scientists are like artists; they reach reality in a series of miracles, called creative leaps. Chapter 4 shows that this is a caricature of a solution, even for the arts. It is not needed either: treating science as part of history removes the problems the idea of individual creativity was designed to solve.

It also removes the apparent difference between the sciences and the arts. Consequences for the idea of progress are explained in chapter 5: judgements of progress are relative judgements in both.

Some readers who have followed me up to the end of chapter 4 will point out that while my criticisms of Reason and Rationality may be correct for older versions, they no longer apply to Popper's 'critical rationalism'. Chapter 6 answers this objection. It shows that a universally critical philosophy like Popper's either has no substance – it excludes nothing – or it removes ideas and blocks actions we might want to retain. Empty verbiage or stumbling block – these are the alternatives open to Popper, and he uses now the one, now the other, depending on the kind of criticism he wants to evade. For example he is not averse to applying 'some form of imperialism' against people who resist entering the wonderful palace of Western Civilization (*The Open Society and Its Enemies* Vol. 1, New York 1963, p. 181).

Are there better philosophies? Are there philosophies of science which provide understanding without removing ideas and actions that might advance knowledge? There are such philosophies – and chapter 7 gives an example: the philosophy of Ernst Mach. Ernst Mach made contributions to physics, physiology, the history of science, the history of ideas and to general philosophy. He had no difficulty pursuing such a wide range of interests for he lived and worked before the Vienna Circle had redefined and drastically narrowed our image of the sciences. Science, for Mach, was a historical tradition. He used stories, not abstract models, to explain its development and to prepare scientists for their task. He searched for, and later applauded, relativistic theories of space and time (the introduction to the *Physical Optics* which is severely critical of the special theory of relativity now seems to be a fake concocted by Mach's son, Ludwig), and he anticipated basic features of quantum mechanics. Most importantly, however, he gave an account of theory-construction that combined historical, theoretical and psychological considerations and provided a model for Einstein's methods of research. Speaking the language of dialectical materialism we may say that Mach gave a materialistic account of the growth of (scientific) knowledge. This is what Mach *did*.

What he is *said to have done* is another story. For most historians and philosophers Mach was a narrowminded positivist who wanted science to stick to simple observations and who rejected atoms and relativity as being too general and too abstract. The difference is not the result of profound difficulties in the interpretation of obscure and puzzling texts – Mach's views are expressed clearly, in simple terms, and they can be found in all his major works. It can only be explained by considerable carelessness on the part of almost all his critics. Thus a study of Mach not only introduces us to a wonderful person and a fascinating philosophy of science; it also teaches us an interesting lesson about the nature of scholarship: 'experts' frequently do not know what they are talking about and 'scholarly opinion', more often than not, is but uninformed gossip.

Mach was not the only victim of ideologically motivated ignorance. Aristotle, leading Church theoreticians at the time of Galileo, and Niels Bohr in our century are other examples. I dealt with Niels Bohr in Vol. 1, chapter 16 of my *Philosophical Papers* (Cambridge, Mass. 1981). Chapter 8 shows that the view of the continuum held by scientists from Galileo (without hesitation) to Weyl (with much hesitation) was a step back when compared with Aristotle's account. Chapter 9 analyses an often quoted and much discussed letter – Bellarmino's letter to Foscarini – in the light of an ancient debate about the authority of expert knowledge. It shows that the position of the Church was stronger and more humane than is generally assumed.

Chapter 10 discusses the difficulties the phenomenon of incommensurability creates for theoretical traditions (and for Hilary Putnam, one of its defenders). Chapter 11 is part of my contribution to a debate that was initiated and published by the Columbia School of Architecture. The Statement that initiated the debate bemoans the alleged 'chaos' of modern philosophical thought and calls for a unified ideology. I contest both the diagnosis (there may be increasing chaos in philosophy departments, but there certainly is increasing uniformity in the world) and the solution. My main point is that *collaboration does not need a shared ideology*.

Chapter 12, finally, contains a summary of 'my' philosophy (which, of course, is not mine, but a condensation of reasonable ideas from all over the world) surrounded by answers to criticisms. It was written, in German, for a collection of forty essays praising or condemning or yawning about my work (*Versuchungen*, Hans Peter Duerr ed. 2 Vols., Frankfurt 1980, 1981) and was translated and rewritten for this volume. The chapter will make it clear that my concern is neither rationality, nor science, nor freedom – abstractions such as these have done more harm than good – but the quality of the lives of individuals. This quality must be known by personal experience before any suggestions for change can be made. In other words: suggestions for change should come from friends, not from distant 'thinkers'. It is time to stop ratiocinating about the lives of people one has never seen, it is time to give up the belief that 'humanity' (what a pretentious generalization!) can be saved by groups of people shooting the breeze in well heated offices, it is time to become modest and to approach those who are supposed to profit from one's ideas as an ignoramus in need of instruction, or, if business is concerned, as a begger and not as heaven's greatest gift to the Poor, the Sick and the Ignorant.

The title of chapter 12 which is also the title of the book means two things: some thinkers, having been confused and shaken by the complexities of history, have said farewell to reason and replaced it by a caricature; not being able to forget tradition (and not being averse to a little PR), they have continued calling this caricature reason (or Reason, with a capital R, to use my own terminology). Reason has been a great success among philosophers who dislike complexity and among politicians (technologists, bankers etc. etc.) who don't mind adding a little class to their struggle for world domination. It is a disaster for the rest, i.e. practically all of us. It is time we bid it farewell.

The essays that led to these chapters were written for different occasions, they differed in style and they occasionally overlapped. Some essays were rather scholastic, others came from informal talks, still others were reactions to inquiries and comments. I have rewritten most of them, but I have retained their stylistic diversity and some of the repetitions. My beautiful, good and very patient friend Grazia Borrini was kind enough to work her way through a variety of versions. It is through her that I became acquainted with the vast subject of 'development'. Without her gentle but firm criticism this book would be less well argued, more abstract and certainly much more obscure than – alas! – it still is.

1
Notes on Relativism

Encountering unfamiliar races, cultures, customs, points of view, people react in various ways. They may be surprised, curious and eager to learn; they may feel contempt and a natural sense of superiority; they may show aversion and plain hatred. Being equipped with a brain and a mouth they not only feel, they also talk – they articulate their emotions and try to justify them. Relativism is one of the views that emerged from this process. It is an attempt to make sense of the phenomenon of cultural variety.

Relativism has a long history; it goes back at least as far as the late Bronze Age in the Near East, a period the Egyptologist J. Henry Breasted called the 'First Internationalism'. It was discussed and turned into a doctrine by the Greeks, during the transition from the matter-oriented cosmologies of the Presocratics to the political views of the Sophists, Plato and Aristotle. It inspired the sceptical movement and through it the predecessors of the Enlightenment such as Montaigne and the interpreters of the travel reports of the 16th and 17th centuries. It continued throughout the Enlightenment and it is quite fashionable today, as a weapon against intellectual tyranny and as a means of debunking science. Relativistic ideas and practices are not restricted to the West and they are not an intellectual luxury. They occurred in China and they were developed into a fine art by African Natives after an encounter with different races, customs and religions had shown them the many ways of living that exist on this earth.[1]

The wide distribution of relativism makes it a difficult topic to discuss. Different cultures emphazise different aspects and express them in ways most suited to their interests. There are simple versions from which we all can learn and sophisticated versions which are for specialists only. Some versions are based on a feeling, or an attitude, others resemble answers to mathematical problems. Occasionally there is not even a version; there is just a word – 'relativism' – and a (loving, or angry, but at any rate longwinded) reaction to it. To deal with this abundance I shall relinquish the unity suggested by the one word 'relativism', and discuss a variety of points of view instead. I start with some practical observations.

1 Practical Relativism (Opportunism)

Practical relativism (which overlaps with opportunism) concerns the manner in which views, customs, traditions different from our own may affect our lives. It has a 'factual' part dealing with how we *can* be affected and a 'normative' part dealing with how we *should* be affected (how the institutions of a state should deal with cultural variety). To discuss it, I introduce the following *thesis*:

> R1: individuals, groups, entire civilizations may profit from studying alien cultures, institutions, ideas, no matter how strong the traditions that support their own views (no matter how strong the arguments that support these views). For example, Roman Catholics may profit from studying Buddhism, physicians may profit from a study of the *Nei Ching* or from an encounter with African witch doctors, psychologists may profit from a study of the ways in which novelists and

1. 'The lack of prejudice in the Native', writes Karen Blixen about her experience in Kenya (Isak Dinesen, *Out of Africa*, New York 1972, p. 54), 'is a striking thing for you expect to find dark taboos in primitive people. It is due, I believe, to their acquaintance with a variety of races and tribes, and to the lively human intercourse that was brought upon East Africa, first by the old traders of ivory and slaves, and in our days [the Thirties] by the settlers and big-game hunters. Nearly every native, down to the little herd boy of the plains, had in his day stood face to face with a whole range of nations as different from one another, and to him, as a Sicilian to an Esquimo: Englishmen, Jews, Boers, Arabs, Somali, Indians, Swaheli, Masai and Kawirondo. As far as receptivity of ideas goes, the Native is more a man of the world than the suburban or provincial settler or missionary, who has grown up in a uniform community and with a set of stable ideas. Much of the misunderstanding between white people and the Natives arises from this fact.'

actors build a character, scientists in general may profit from a study of unscientific methods and points of view and Western civilization as a whole can learn a lot from the beliefs, habits, institutions of 'primitive' people.

Note that R1 does not recommend the study of unfamiliar institutions and views and it certainly does not turn such study into a methodological requirement. It only points out that the study may have effects regarded as beneficial by the defenders of the status quo. Note also that not all people who allow alien views and customs to influence their outlook formulate theses about the process. They may act as they do because of a feeling of trust towards human beings, or because they are still in touch with the rest of nature (people have learned from animals and plants as well as from humans), or because of strong imitative tendencies. Concentrating on a *thesis* (such as R1) therefore already restricts the discussion; it assumes that the parties put their motives into words and that they use these words instead of relying on examples, empathy, magic or other non-verbal means.

There exists a wide spectrum of responses to R1, the following four among them.

A. The thesis is rejected. This happens when a tightly knit world view that reaches down into the daily lives of the believers is regarded as the only acceptable measure of truth and excellence. The laws of Deuteronomy, Plato's perfect state, Calvin's Geneva, some 20th century cults are examples. Many scientist would like their ideas, products and world views to achieve comparable eminence[2] – and they are getting very close to having their wishes fulfilled.'[3]

B. The thesis is rejected, but only in certain areas. This occurs in pluralistic cultures containing weakly interacting parts (religion, politics, art, science, private and public actions etc.), each part being guided by a well defined and exclusive paradigm. Individuals are cut up accordingly: 'as a Christian' a person may rely on faith, 'as a scientist' (s)he must use evidence. Or as a historian said of Calvin, commenting on the execution of Servetus: 'As a man he was not cruel, but as a theologian he was merciless; and it was as a theologian that he dealt with Servetus.'[4]

A still more liberal response, **C**, encourages an exchange of ideas and attitudes between different domains (cultures) but subjects them to the laws that rule the domain (culture) entered. Thus some medical researchers acknowledge the usefulness of non-Western medical ideas and therapies but add that such things were discovered by scientific means and must be confirmed with their help; they do not have any independent authority.

Finally, on the extreme 'left' of our spectrum we have the view, **D**, that even our most basic assumptions, our most solid beliefs, and our most conclusive arguments can be changed—improved, or defused, or shown to be irrelevant — by a comparison with what at first looks like undiluted madness.

A to **D** (and other responses) have played an important role in the history of the human race; the fate of freedom, toleration and

2. In his book *On Human Nature*, Cambridge, Mass 1978, pp. 192f, Edward O. Wilson writes as follows: ' . . . religion . . . will endure for a long time as a vital force in society. Like the mythical giant Antaeus who drew energy from his mother, the earth, religion cannot be defeated by those who merely cast it down. The spiritual weakness of scientific naturalism is due to the fact that it has no such primal source of power. While explaining the biological sources of religious emotional strength [a bold claim, not substantiated by Wilson's research-report], it is unable in its present form to draw on them, because the evolutionary epic denies immortality to the individual and divine privilege to the society [this is of course true – but people can and do live full lives without these ingredients; absence of divine privilege does not mean absence of reverence and spiritual fulfillment; materialism does] and it suggests only an existential meaning for the human species [it does?]. Humanists will never enjoy the hot pleasure of spiritual conversion and self-surrender; scientists cannot in all honesty serve as priests [yet they are trying hard to usurp this function and the 'self-surrender' to unprejudiced objectivity that goes with it]. So the time has come to ask: does a way exist to divert the power of religion into the services of the great new enterprise [i.e. materialistic science] that lays bare the sources of that power?' In short: does a way exist to make science as powerful as religion used to be and for many people still is?

3. Bakunin predicted 'the reign of scientific intelligence, the most autocratic, despotic, arrogant and elitist of all regimes' (S. Dolgoff, *Bakunin on Anarchy*, New York 1972 p. 319). Writers closer to our own age confirm the prediction. Thus Daniel Bell believes that 'the entire complex of social prestige will be rooted in the intellectual and scientific communities' ('Notes on the Postindustrial Society I', in *The Public Interest*, Winter 1967) while J.K. Galbraith holds that 'power in economic life has over time passed from its ancient association with land to association with capital and then on, in recent times, to the composite of knowledge and skills that comprises the technostructure' (*The New Industrial State*, Boston 1967). The 'intellectual and scientific communities' increasingly reject outside interference – they are best qualified, they say, to distinguish between important and unimportant matters and to recognize the limits of their achievements. Leading scientists and philosophical opportunists eager to be on the right side claim in addition that the scientific spirit is capable of policing not only science, but our entire existence. Cf. E.O. Wilson, *op. cit.*, B.F. Skinner, *Beyond Freedom and Dignity*, New York 1971; and, on the whole problem, Noam Chomsky, 'Intellectuals and the State', in *Towards a New Cold War*, New York 1986, pp. 60ff.

4. Robert Jay Lifton has examined this phenomenon of 'doubling' in the extreme case of deathcamp physicians: *The Nazi Doctors*, New York 1986 and *The Future of Immortality*, New York 1987, esp. chapter 14. The phenomenon is more widespread than he indicates. Four of his five characteristics (*Future*, p. 196) apply to many biologists and social scientists, all five apply to medical researchers who torture animals so that knowledge may increase and humans, perhaps, live a little longer. 'Within these struggles around professional identity', writes Lifton (*Future*, p. 91), discussing Auschwitz physicians, 'the strongest single pattern was the technicality of everything. As an SS doctor said to me: "Ethics was not a word used in Auschwitz. Doctors and others spoke only about how to do things most efficiently, about what worked best." 'It does not need an SS officer to put efficiency or technological 'sweetness' above human concerns. When it was discovered, in November 1944, that the Germans could not possibly manufacture an atomic bomb 'the idea began arising among the atomic physicists [engaged in building an American bomb] that the bomb was no longer needed, and that humanity could be spared the apocalypse which they had been preparing for it. Nevertheless there were not many who urged immediate cessation of work on the bomb. Such a sweeping act of renunciation, at the very moment when success was in sight, did not come easily to them who for many months had given their best to bring the project to realization' (M. Rouzé, *Robert Oppenheimer*, New York 1965, p. 68). Details are found in Richard Rhodes' comprehensive study *The Making of the Atomic Bomb*, New York 1986. Cf. for example Oppenheimer's speech (p. 761) with its Hegelian progressivism. The book shows that some scientists, Bohr and Szilard among them, realised that atomic weapons made traditional political thought obsolete and suggested alternatives. Bohr was aware of the danger but also of the possibility that the very size of the danger might lead to greater political openness, worldwide (this Rhodes calls the 'complementarity of the bomb'). Today many scientists work actively towards nuclear disarmament. But there are other scientists for whom research comes before détente. Thus the director of the Lawrence Livermore National Laboratory arguing (in a letter to Congress, reprinted in the *Bulletin of the Atomic Scientists*, November 1985, p. 13) against a comprehensive testban pointed out that 'weapons design experts would invariably leave the weapons program because they could not verify their theoretical ideas with experiments': the need of scientists to continue playing the game of science in the simplest and most efficient manner overrules questions of peace and survival. Similarly, objections to Star Wars were answered by the remark that human knowledge must increase. The special group at the Livermore Laboratory dealing with Star Wars consisted of 'bright young hot shots who are socially maladjusted. All their time and energy is spent on science. There are no women, no outside interests. They focus on far-out technical problems . . . ' (Hugh deWitt as quoted in W. Broad, *Star Warriors*, New York 1985, p. 25). Western civilization as a whole now values efficiency to an extent that occasionally makes ethical objections seem 'naive' and 'unscientific'. There are many similarities between this civilization and the 'spirit of Auschwitz'. I would not conclude, however (as does Ashis Nandi in his M.N. Roy memorial lecture of 1980 – abbreviated version published as 'The Pathology of Objectivity', *The Ecologist* 1981), that prominent features of this civilization are pathological. This is hardly more intelligent and certainly not more enlightening than the charge of 'irrationaliy' or 'lack of a scientific basis' used so freely by the bigshots of the status quo. Western civilization (and, for that matter, Auschwitz) is one of the many possible manifestations of human life and the problems it creates are not solved by calling it names).

rationality are inextricably intertwined with the way in which influential groups and entire cultures have dealt with diversity (of ideas, customs, attitudes) and thus with R1. In this section and the next I shall only examine the claims of *science* and of science-based developments. By 'science' I mean modern natural and social science (theoretical and applied) as interpreted by most scientists and a large section of the educated public: an inquiry that aims at objectivity, uses observation (experiment) and compelling reasons to establish its results and is guided by well defined and logically acceptable rules. I shall argue that neither values, nor facts, nor methods can support the claim that science and science-based technologies (IQ tests, science-based medicine and agriculture, functional architecture, and the like) overrule all other enterprises.

Speaking of *values* is a roundabout way of describing the kind of life one wants to lead or thinks one should lead. Now people have arranged their lives in many different ways. It is therefore to be expected that actions that seem perfectly normal in one culture are rejected and condemned in another. To take an example (an actual case I heard from Christina von Weizsäcker): a physician suggests X-rays to pinpoint the illness of the member of a Central African tribe. His patient wants him to use other methods: 'what is going on in my inside is nobody's business'. Here the desire to know and, on the basis of knowledge, to cure in the most efficient way possible clashes with the wish to maintain privacy and the integrity of the body (of the person). Arguing about values means examining and resolving conflicts of this kind.

Is the wish of the patient a reasonable wish? It is reasonable for a community that values privacy and the integrity of the body and expects its wise men to work within the limits defined by these values.[5] It is unreasonable for a community where efficiency and the pursuit of knowledge overrule everything. (Large parts of Western civilization seem to function in this manner – cf. footnote 4). It is unreasonable but tolerated in a society that welcomes efficiency and the rule of experts but makes room for personal idiosyncracies. It is unreasonable and denounced when-

5. It seems that early support for acupuncture in China came from groups that objected to anatomy and intrusive methods of healing and diagnosis because they regarded the human body as sacred: Ilza Veith, ed., *The Yellow Emperor's Classic of Internal Medicine*, Berkeley and Los Angeles 1966, pp. 2f. (Joseph Needham is inclined to deny the existence of such tendencies). For early Western ideas on dissection cf. C.D. O'Malley, *Andreas Vesalius of Brussels*, Univ. of California Press 1965, ch. 1.

ever the wish defies established social rules. It is both reasonable and unreasonable (or whatever other words are used to indicate agreement or conflict with basic demands) in a society that encourages the development, within a single framework, of many different ways of living. Some will accept the wish and encourage it, others heap abuse and ridicule on its proponents. The debates about abortion, euthanasia, gene manipulation, artificial insemination and the (intellectual, political, economic, military) exchanges between different cultures illustrate the way in which values influence opinions, attitudes and actions. Many debates continue even after the contestants have received all the available information and have started arguing in the same way. The tensions that remain are tensions between values, not between good and bad or between complete and insufficient information (though many debates are complicated by these elements as well), and not between reason and irrationality (though the values defended are often made part of reason).[6]

There are essentially three ways to resolve such tensions: power, theory, and an open exchange between the colliding groups.

The way of power is simple and quite popular. There is no argument; there is no attempt to understand; the form of life that has the power imposes its rule and eliminates behaviour contrary to it. Foreign conquests, colonization, developmental programmes and a large part of Western education are examples.

The theoretical approach does use understanding, but not the understanding of the parties concerned. Special groups, philosophers and scientists among them, study the conflicting values, arrange them in systems, provide guidelines for the resolution of conflicts – and that settles the matter. The theoretical approach is conceited, ignorant, superficial, incomplete and dishonest.

It is conceited because it takes it for granted that only intellectuals have worthwhile ideas and that the only obstacle to a harmonious world is dissension in their ranks. Thus Roger Sperry in an interesting and challenging book[7] observes that 'current world conditions call for a unified global approach with

6. Separating facts, values and rationality is of course an artifice. Facts are constituted by procedures that contain values, values change under the impact of facts and principles of reasoning assume a certain world order (the law of non-contradiction is absurd in an absurd world). I use the artifice to simplify the discussion. This restricts my arguments and should therefore be acceptable to the defenders of objective values.

7. *Science and Moral Priority*, Westport, Conn. 1985. The quotations are from pages 72, 6, 32, 75.

value perspectives . . . that will include the welfare of the total biosphere.' At present, Sperry says, such a unified approach is blocked by 'the crisis of contemporary culture', namely 'the profound contradiction between the traditional humanist views of man and the world, and the value devoid mechanistic descriptions of science.' To remove the contradiction Sperry suggests a reform of science that eliminates reductionism and 'puts mind and consciousness in the driver's seat.' The resulting world view still differs from the various 'mythological, intuitive, mystical or otherworldly frames of reference by which man has . . . tried to live and find meaning.' There still exist 'profound contradiction[s]' between 'the natural cosmos of science' and cultures outside the realm of Western civilization. But these contradictions do not add up to a 'crisis' and science is not changed to resolve them: cultures outside the sciences and the humanities simply do not count. Many leading intellectuals think along similar lines.[8]

The theoretical approach is, secondly, ignorant. For example, it overlooks that many of the problems now faced by Third World countries (hunger, overpopulation, spiritual decay) arose because ecologically sound and spiritually satisfying forms of life were disrupted and replaced by the artifices of Western Civilization.[9] The 'different mythological, intuitive, mystical or otherworldly frames of reference' mentioned by Sperry were not just dream castles; they delivered what they promised; they ensured material survival and spiritual fulfillment in the most adverse circumstances.[10] The messengers of progress and civilization

8. Thus E.O. Wilson, *op. cit.*, welcomes the stimulation of knowledge caused by the appearance of 'anti-disciplines' (p. 8) to established sciences. The humanities (which are well-established, academically) are accepted as anti-disciplines. Non-academic ideas and points of view are not.

9. Cf. the literature in fn. 3 and 4 of the Introduction. Non-scientific cultures may also have a clearer conception of the dangers of radically new knowledge. Numerous myths tell us what intellectuals have realised only recently and not without putting up a fight — that information separated from the circumstances of its origin has destructive tendencies, and that the course of nature cannot be changed without repercussions.

10. I do not mean to say that all is well in indigenous societies and that outside help is never needed. Parasites, infectious diseases, congenital deficiencies constitute enormous problems, some of which were alleviated by Western medicine (details in W.W. Spink, *Infectious Diseases*, Minneapolis 1978 – a somewhat too optimistic book). There is no perfect society just as there is no perfect human body. However the authors I am criticizing go much further. They not only assume that help may be needed, they take it for granted that *any* change in the direction of Western civilization and especially of Western science is bound to be an improvement. This is simply not true.

destroyed what they had not built and ridiculed what they did not understand. It would be shortsighted to assume that they alone now possess the keys to survival.

Thirdly, the theoretical approach is surprisingly superficial. It replaces the rich complex of ideas, perceptions, actions, attitudes, and gestures, down to the most fleeting smile of the tiniest baby that arises from the working of a particular value, by arid and abstract concepts and assumes that a 'rational' choice between these chimaeras already decides the matter: '. . . . the deep epistemological problems that face anyone who wishes to describe 'human nature' seem not to have been taken into account by . . . theorists. Faced with the extraordinary richness and complexity of human social life in the past and present, they have chosen the nineteenth-century path of describing the whole of humankind as a transformation of European bourgeois society.'[11] The superficiality applies also in approaches to the sciences. There is little discussion of the great variety of scientific disciplines, schools, approaches, answers. All we get is a monolithic monster, 'science', that is said to follow a single path and to speak with a single voice.

The theoretical approach is, fourth, incomplete: it remains silent about the question of enforcement. This does not mean that the theoreticians have no opinion on the matter. They have very definite opinions. They hope that their suggestions will eventually be accepted by the institutions of Western industrial states and will from there seep first into education and then into development. Like their predecessors, the colonial officials, they have no compunction about letting power enforce their ideas. But unlike colonial officials they do not apply the power themselves; on the contrary, they emphazise rationality, objectivity and tolerance, which means that they are not only disrespectful, ignorant, and superficial but also quite dishonest. Fortunately there are now scientists who, guided by a deep respect for all forms of human existence, have discovered the inherent strength of 'primitive' views and 'archaic' institutions and have changed their views of knowledge accordingly. As they see it, research is not a privilege of special groups and (scientific) knowledge not a

11. R.C. Lewontin, Steven Rose and Leon J. Kamin, *Not in Our Genes*, New York 1984, p. 245. The authors criticize sociobiology; they criticize it for its superficiality, not because it puts theory in the place of individual decisions. But even the most complex theory can only narrate what a particular set of values has done in the past, it cannot predict what the values will do in new and unforeseen circumstances. To find future results we have to turn to those using the values, i.e. we have to ask them to make their own decisions.

universal measure of human excellence. Knowledge is a local commodity designed to satisfy local needs and to solve local problems; it can be changed from the outside, but only after extended consultations that include the opinions of all concerned parties. Orthodox 'science', in this view, is one institution among many, not the one and only repository of sound information. People may consult it; they may accept and use scientific suggestions – but not without having considered local alternatives and certainly not as a matter of course.[12] The new forms of knowledge that have emerged from this approach are less superficial and better adapted to the needs of the modern world than the procedures and results of orthodox science.[13]

The remarks just made show that values affect not only the *application* of knowledge but are essential ingredients of *knowledge itself*. Much of what we know about people, their habits, idiosyncracies, and prejudices, arises from interactions (between people) that are shaped by social customs and individual preferences; this knowledge is 'subjective' and 'relative'. It is preferable to the 'knowledge' that comes from an interaction between people and experimental arrangements (psychological tests; genetic studies; theories of cognition), for it sustains personal contacts instead of undermining them. There exist of course areas (materials science, for example) where quantitative experimental information seems to defeat all rivals. But the victory is not an 'objective fact' (like a military victory it depends on the aims of those engaged in the battle; in the example the aim is technological improvement as defined at a certain phase of Western civilization), it is not a matter of course (it must be established – it cannot be taken for granted – and it has to be judged by those who are supposed to profit from it[14]), the success cannot be extrapolated (the fact that experiments have advanced

12. The procedure, like any procedure, has exceptions. The occurrence of widespread and quickly expanding diseases may require quick and tyrannical action on the part of those who have the power and think they have the knowledge to deal with the emergency. My point is that such cases should be treated as exceptions. Local consultation should be carried as far as possible and should be resumed the moment the danger recedes.

Suppression and mass killings are other examples where interference may be needed. But the presumptive saviours must realise that they can only rely on their own firm convictions and that no 'objective' values will come to their rescue when their efforts fail, or worsen matters, or are condemned by the moral consensus of later generations. We condemn Auschwitz and some of us condemn Truman's decision to drop the atom bomb on Hiroshima and Nagasaki because this is the kind of people we are and not because we have a direct line to heaven.

some parts of physics says nothing about their role in psychology or in other parts or periods of physics) it changes with time (there were times when qualitative information about materials far exceeded quantitative knowledge in content and technological efficiency – see footnotes 38, 39 and text), and even efficient knowledge may be rejected because of the way its acquisition disturbs important social values. 'Is it not possible', asked Kierkegaard (*Papirer,* ed. Heiberg, vii pt, i, sec. A, Nr 182), 'that my activity as an objective observer of nature will weaken my strength as a human being?' All this means that (criteria of) success and acceptance change from case to case, and in accordance with the values of those interested in a particular area of knowledge.

13. In *Science in a Free Society*, London 1978, p. 29, I called an exchange of the kind described in the text an 'open exchange'. 'There are . . . at least two different ways of collectively deciding an issue', I wrote on that occasion, 'which I shall call a guided exchange and an open exchange respectively. In the first case some or all participants adopt a well specified tradition and accept only those responses that correspond to its standards . . . A rational debate is a special case of a guided exchange . . . An open exchange, on the other hand, is guided by a pragmatic philosophy. The tradition adopted by the parties is unspecified in the beginning and develops as the exchange goes along. The participants get immersed into each others' ways of thinking, feeling, perceiving to such an extent that their ideas, perceptions, world views may be entirely changed – they become different people participating in a new and different tradition. An open exchange respects the partner whether he is an individual, or an entire culture while a rational exchange promises respect only within the framework of a rational debate. An open exchange has no organon though it may invent one, there is no logic though new forms of logic may emerge in its course.'

Some objectivist philosophers have by now come rather close to this point of view. Thus Habermas admits (*Habermas, Authority and Solidarity*, P. Dews, ed., London 1986, p. 205) that when trying to design 'just institutions for a certain type of society under given historical circumstances', a philosopher must join the citizens and 'move within the horizon of a shared tradition from the start'. Also (p. 160), 'the moral philosopher must leave the substantive questions [of moral discourse] to the participants . . . or tailor the cognitive claims of normative theory from the outset to the role of the participant.' However, he still thinks that there exists a 'universal core of the moral point of view' (p. 205) which can be revealed by a 'critique of value scepticism and value relativism' (p. 161) and which is more than an accidental temporary overlap of cultures. Needless to say, no practical debate about colonialism, development, or armed interference will be affected by this belief – and so we may leave it undisturbed in the hope that the inner philosophical quarrels created by it will be big enough to keep philosophers from interfering in more substantial matters.

14. Scientifically satisfying musical instruments or rooms often sounded terrible to the ears of music lovers and had to be improved by artisans. Today, of course, the technological sound has itself become a measure of excellence.

To sum up: decisions concerning the value and the use of science are not scientific decisions; they are what one might call 'existential' decisions; they are decisions to live, think, feel, behave in a certain way. Many people never made decisions of this kind; many are now forced into them: people in 'developing nations' became doubtful about the blessings of Western ways, citizens of Western countries look with suspicion at the products of technology that have arisen in their midst (this is written in the aftermath of the Chernobyl accident of April 1986). The 'existential' nature of the decisions for or against a scientific culture is the main reason why the *products* of science (TV sets; atomic bombs; penicillin) are not ultimately decisive. They are good or bad, helpful or destructive, *depending on what kind of life one wants to live.*

My second comment concerns *facts*. It is not at all certain that even a scientific comparison, based on scientific values, or scientific and non-scientific cultures will always favour the former. Agreed – there will be advantages in large areas of abstract knowledge and practical skills. But there are other areas where the superiority of a scientific-technological approach is far from obvious. Thus scholars studying the history of non-Western civilizations and communities found that hunger, violence, increasing scarcity of goods and services that had once been available in abundance, alienation, and 'underdevelopment' can often be traced to the disruption, due to the advance of Western science and technology, of complex, fragile but surprisingly successful socioecological systems.[15] Or assume that a sufficient number of sick people is classified in accordance with the most advanced methods of Western medicine and then divided into two groups; one group is treated in the accepted Western way (assuming there is one such way), the other by representatives of a non-scientific form of medicine such as acupuncture. The results, too, are examined by Western doctors (follow-up studies over years included, where necessary). What will the outcome be? Will Western medicine always have better results, according to Western standards? Will it have better results in the majority of cases? Will there be areas where it fails and where other methods succeed? The answer to all these questions is: we do not know.

Nobody will deny that there have been great and surprising

15. In addition to the literature given in footnote 9, consult M. Watts, *Silent Violence*, Berkeley and Los Angeles 1983.

successes and that they were achieved by a combination of scientific materialism and sometimes simple, sometimes rather sophisticated experimental techniques. But these are isolated events which do not yet establish the universal usefulness of the combination and the universal failure of all existing alternatives. We simply don't have an overall picture based on evidence rather than on unchecked generalizations. Add to this that the care of the aged, the treatment of the insane, and the education of small children (their emotional education included), all of which belong to the area of well being, are left to experts in industrial societies but are matters of family or community action elsewhere; consider also that there is no unambiguous criterion of health – well being is evaluated differently at different times and in different cultures – and it will become clear that the question of the comparative excellence of scientific and non-scientific procedures has never been examined in a truly scientific way. Once again the apostles of science are found to lack the scientific credentials for their faith. This is not an indictment of science; it only shows again that the choice of science over other forms of life is not a scientific choice.

Non scientific forms of life *were* examined, says a popular counter-argument, because scientists checked and eliminated the alternatives when they first met them. For example, Indian remedies (which were popular with nineteenth-century medical practitioners in the United States) disappeared when the pharmaceutical industry provided superior substitutes.

This counter-argument is both incorrect and irrelevant. It is incorrect because many so-called victories of science-based practices were not results of systematic comparative research, but of anectodal evidence enlarged by independent social developments, political (institutional) pressures, and powerplay. Take again the example of medicine. According to Paul Starr (*The Social Transformation of American Medicine*, New York 1982), important changes in the role of doctors, the transition to a more impersonal (or, in technical terms, 'objective') approach included, were largely due to social developments and not to advances in medical knowledge. The developments affected what was regarded as correct medical procedure and created the appearance of progress without corresponding research. Stanley Joe Reiser (*Medicine and the Rise of Technology*, Cambridge 1978) discusses the role of new technologies in a similar way. The idea that instruments are better than human observers fitted into a general trend towards impersonality, and diagnoses based on personal contact were thus already regarded as inappropriate. Improvements in health were often due to more and better food, sanitation, improved working conditions, a treatment-independent periodicity of major diseases and not to improved medical practice (details in R.H. Shryock, *The Development of Modern Medicine*, University of Wisconsin Press, Madison, Wisconsin 1979, pp. 319ff).

Considering that writers such as Lewis Thomas and Peter Medawar put the beginning of scientific medicine into the nineteen thirties (cf. *The London Review of Books*, 12 Feb – 2 March 1983, pp. 3ff), we are led either to infer that, seen from a sternly scientific point of view, pre-twentieth-century medicine had reputation without content and progress without substance, or to admit that medicine can be successful without being scientific. 'Medicine,' writes Lewis Thomas (*The Youngest Science*, New York 1983, p. 29), 'for all its façade as a learned profession, was in real life a profoundly ignorant occupation.' The speedy acceptance of prefrontal and transorbital lobotomy by large sections of the medical establishment in the absence of any even half way decent evidence about results shows that professional acceptance still does not mean excellence and that the argument from professional agreement must be used with great caution (see Elliot S. Valenstein, *Great and Desperate Cures*, New York 1986). Numerous procedures that were announced with a flourish and imposed on an unwary public (use of calomel in the 19th century; irradiation of enlarged thyroid glands in children only ten years ago; the Halstead method) were fashions without proper empirical backing. Now I do not deny that part of the reputation of doctors rests on genuine and often very surprising successes of medical research (Thomas, p. 35). The impact of sulfanilamide, of penicillin, of effective new methods of antenatal diagnosis can hardly be exaggerated. But there are other phenomena which forbid an unqualified inference from such isolated successes to the complete and hopeless invalidity of non-orthodox forms of medicine. *Every case must be examined separately and judged on its own merits*, independently of the practical confidence and the theoretical fashions of the time.

The counter-argument is also irrelevant: every genuine scientific victory is achieved by using a variety of weapons (instruments, concepts, arguments, basic assumptions). But the weapons change with the advance of knowledge. A repetition of the contest therefore may, and often does, have a different outcome; victory turns into defeat and vice versa. Many once utterly ridiculous views are now solid parts of our knowledge. Thus the idea that the earth moves was rejected in antiquity because it clashed with facts and the best theory of motion then available; a recheck, based on a different, less empirical and at the time highly speculative dynamics convinced scientists that it had been correct after all. It convinced them because they were not as adverse to speculation as their Aristotelian predecessors had been. The atomic theory was frequently attacked, both for theoretical and for empirical reasons; during the second third of the 19th century some scientists thought it hopelessly outdated; yet it was revived by ingenious arguments and it is now a basis of physics, chemistry and biology. The history of science is full of theories which were pronounced dead, then resurrected, then pronounced dead again only to celebrate another triumphant comeback. It makes sense to preserve faulty points of view for possible future use. The history of ideas, methods and prejudices is an important part of the ongoing practice of science and this practice can change direction in surprising ways.

What is true of theories applies with even greater force to the applied sciences and science-based arts like medicine. In medicine we have not only recurrent fashions (the fashion of a more 'personal' approach; the fashion of therapeutic nihilism that played a large role in antiquity and returned around the turn of the century) and perennial fluctuations between alternatives (example: 'an illness is a local disturbance that must be removed' versus 'an illness is the body's way of overcoming disturbances and should be supported'), we have also science-independent changes of reputation and of basic terms such as 'health' and 'well being'. Medicine can only profit from making history part of its practice and its research.

John Stuart Mill, in his immortal essay *On Liberty*,[16] went still further. He advised researchers not only to retain ideas that had been tested and found wanting but to consider new and untested conceptions as well, no matter how absurd their first appearance. He gave two reasons for his suggestion: a variety of views, he said, is needed for the production of 'well-developed *human beings*'; and it is needed for the improvement of *civilization*:

> What has made the European family of nations an improving, instead of a stationary, portion of mankind? Not any superior excellence in

16. I quote from M. Cohen, ed., *The Philosophy of John Stuart Mill*, New York 1961, pp. 258, 268f, 245f.

them which, when it exists, exists as the effect, not as the cause, but their remarkable diversity of character and culture. Individuals, classes, nations have been extremely unlike one another: they have struck out a great variety of paths, each leading to something valuable; and although at every period those who travelled in different paths have been intolerant of one another, and each would have thought it an excellent thing if all the rest would have been compelled to travel his road, their attempts to thwart each other's development have rarely had any permanent success, and each has in time endured to receive the good which the others have offered. Europe is, in my judgement, wholly indebted to this plurality of paths for its progressive and many sided development.

According to Mill a plurality of views is also needed *in the sciences* – 'on four different grounds'. First, because a view one may have reason to reject may still be true. 'To deny this is to assume our own infallibility'. Secondly, because a problematic view 'may and very commonly does, contain a portion of truth; and since the general and prevailing opinion on any subject is rarely 'or never the whole truth, it is only by the collision of adverse opinions that the remainder of the truth has any chance of being supplied.' Thirdly, a point of view that is wholly true but not contested 'will . . . be held in the manner of a prejudice, with little comprehension or feeling of its rational grounds.' Fourthly, one will not even understand its meaning, subscribing to it will become 'a mere formal confession', unless a contrast with other opinions shows wherein this meaning consists.

A fifth, and more technical reason is[17] that decisive evidence against an opinion can often be articulated and found only with the help of an alternative. To forbid the use of alternatives until contrary evidence turns up while still demanding that theories be confronted with facts, therefore, means putting the cart before the horse. And using 'science' to denigrate and perhaps even to eliminate all alternatives means using a well deserved reputation to sustain a dogmatism contrary to the spirit of those who earned it.

17. For details cf. my *Philosophical Papers*, Vol. 1, pp. 144f. The methodological situation I am discussing here undermines statements such as 'the genes hold culture on a leash' (E.O. Wilson, *On Human Nature*. Harvard University Press, Cambridge, Mass. 1978, p. 167) which seem to imply 'objective' limits to human ingenuity. If they do, then this can only be discovered by acting as if the limits did not exist. And there is no cut-off point at which such testing ceases to make 'objective' sense. ('Subjectively', of course, people very soon get tired of barren alternatives.) Cf. my comments on Pavlov in Vol. 1, ch. 6, section 9 of my *Philosophical Papers* and chapter 5, section 3.

Some scientists see science as a steamroller that flattens everything in its path. Thus Peter Medawar (*The Art of the Soluble*, London 1967, p. 114) writes: 'As science advances, particular facts are comprehended within, and therefore in a sense annihilated by, general statements of steadily increasing explanatory power and compass whereupon the facts need no longer be known explicitly. In all sciences we are being progressively relieved of the burden of singular instances, the tyranny of the particular.' But it is just this 'tyranny' or, as I would rather say, this complexity of real life (which is a life among particulars) that keeps our minds flexible and prevents them from being overly impressed by similarities and appearances of lawfulness. Moreover, in the human sciences it would not only be unwise but also immoral and tyrannical to 'annihilate' individual points of view because they do not fit into general frameworks of 'increasing explanatory power'.

S.E. Luria, in a fascinating, informative and often very moving autobiography (*A Slot Machine, A Broken Test Tube*, New York 1985, p. 123), writes as follows: 'What matters in science is the body of findings and generalizations available today: a time-defined cross-section of the process of scientific discovery. I see the advance of science as self-erasing in the sense that only those elements survive that have become part of the active body of knowledge.'[18] 'The model of the DNA molecule worked out by Crick and Watson', continues Luria, 'stands on its own merits. Alternative models have been discarded and forgotten, no matter how vividly they might have been proposed. The . . . story of how the DNA model was achieved, humanly fascinating as it may be, has little relevance to the operational content of science.' Yet it is not this 'operational content' that affects scientists but the way in which it meshes with their personal interests. Luria, for example, prefers events that lead to 'strong inferences', 'predictions that will be strongly supported or sharply rejected by a clear-cut experimental step' (pp. 115f). He 'confess[es] a lack of enthusiasm . . . in the 'big problems' of the universe or of the early Earth or the concentration of carbon dioxide in the upper atmosphere' (p. 119), and he reports that Fermi, for similar reasons, was somewhat cool towards the

18. Piotr Kapitza, while in Cambridge, had the facade of his laboratory adorned by a crocodile. Asked for the meaning of this he replied: 'Well, [this] is the crocodile of science. The crocodile cannot turn its head. Like science it must always go forward with all devouring jaws.' See M. Rouzé, *Robert Oppenheimer*, New York 1965, p, 12.

general theory of relativity (p. 120). A science full of people so inclined will differ considerably from a theoretical science 'loaded with weak inferences' (p. 119); it will also preserve errors: facts based on strong inferences were often undermined or shown to be erroneous (in the sense of those accepting later temporal cross-sections) by chains of weak inferences. (Instances include Galileo's undermining of the arguments against the motion of the earth and Boltzmann's undermining of phenomenological thermodynamics.) The 'operational content' of science at a certain time is therefore the outcome of objective moves made in accordance with subjective interests and interpreted on the basis of assumptions assembled by these interests. We must know the interests to assign the content a proper weight and, perhaps, to correct it. But with this we return to Mill's model.

We may conclude that there exists no scientific argument against using or reviving non-scientific views or scientific views that have been tested and found wanting, but there do exist (plausible, but never conclusive) arguments in favour of a plurality of ideas, unscientific nonsense and refuted bits of scientific knowledge included. This further supports the idea of local knowledge as explained in the text to footnotes 12 and 13.

The third objection to the idea that science overrules all other forms of life comes from the area of *methodology*: the fictitious unit 'science' that is supposed to exclude everything else simply does not exist. Scientists have taken ideas from many different fields, their views have often clashed with commonsense and established doctrines, and they have always adapted their procedures to the task at hand. There is no one 'scientific method', but there is a great deal of opportunism; anything goes – anything, that is, that is liable to advance knowledge as understood by a particular researcher or research tradition.[19] In practice science often oversteps the boundaries some scientists and philosophers try to put in its way and becomes a free and unrestricted inquiry. But such an enquiry cannot deny R1; on the contrary – R1 is one of its most important ingredients. What is exclusive is not science itself but an ideology that isolates some of its parts and hardens them by prejudice and ignorance.

Modern science has gone a long way towards dismantling this ideology. It replaced the 'eternal laws of nature' by historical processes. It created a world view that seems to involve the once 'repugnant' idea of a beginning in time. It has replaced an old, crude and unexamined subject-object distinction by a much more subtle and not easily comprehensible arrangement of facts (complementarity). It has emphasized the need to make subjectivity not only an object, but also an agent of scientific research. Following some ideas of Poincaré, it has introduced qualitative considerations into what seemed to be the most quantitative of all sciences – celestial mechanics.[20] Moreover, it has discovered and studied the marvellous arts, technologies and sciences of cultures and civilizations different from our own.[21] Together with the studies mentioned in footnote 9 (and the vast literature on 'development'), these discoveries show that all nations and not only industrialised countries have achievements from which humanity as a whole can benefit; they make us realise that even the smallest tribe may be able to offer new insights to Western thought; and they have convinced some writers that science and scientific rationalism, far from being one form of life among many, may not even be a form of life.[22] Most importantly for my present purpose, however, they establish that *R1 is not only reasonable* (cf. the above sketch of Mill's arguments) *and a significant part of sciences not inhibited by ideology*, but that it is also *well confirmed*. Or, to express it differently: large parts of science have transgressed the boundaries drawn by a narrow

19. Cf. the references and quotations in my paper 'Was heisst das, wissenschaftlich sein?' *Grenzprobleme der Wissenschaften*, P. Feyerabend und Chr. Thomas, eds., Zürich 1985, pp. 385ff and on p. 188ff. of the present book.

20. The idea of a beginning in time was repugnant because of its closeness to religious creation stories. It was called 'repugnant' by the great astronomer and natural philosopher Arthur Stanley Eddington: *Nature*, Vol. 127 (1931), p. 450. Fred Hoyle, *Facts and Dogmas in Cosmology and Elsewhere*, The Rede Lectures, Cambridge 1982, pp. 2f still opposes 'the modern school of cosmologists who in conformity with Judaeo-Christian theologians believe the whole universe to have been created out of nothing.' As far as I can see the closeness of modern cosmology to creation in the Biblical sense is more imagined than real. Still, this imagined similarity was no obstacle to research.

The idea of complementarity is explained by Niels Bohr in *Atomic Physics and Human Knowledge*, New York 1963. Max Delbrück's lectures *Mind from Matter?*, Blackwell 1985, chs. 17 and 18 discuss its consequences; Peter Fischer, *Licht und Leben*, Konstanz 1985 explains Delbrück's (unsuccessful) search for complementary phenomena in biology; David Bohm, *Wholeness and the Implicate Order*, London 1980 provides an interpretation of his own.

The need to use subjectivity as an instrument of research is emphasized by Konrad Lorenz, *Der Abbau des Menschlichen*, Munich 1983, esp. chapter 4. For historical information on qualitative trends in perturbation theory see J. Moser, *Ann, Math. Stud.*, Nr. 73, Princeton University Press 1973. Historical considerations are found in C.F. von Weizsaecker (*History of Nature*, New York 1964), I. Prigogine (*From Being to Becoming*, New York 1977), H. Haken (*Synergetics*, New York 1983) and in others.

rationalism or 'scientific humanism' and have become inquiries that no longer exclude the ideas and methods of 'uncivilised' and 'unscientific' cultures: *there is no conflict between scientific practice and cultural pluralism.* Conflict arises only when results that might be regarded as local and preliminary and methods that can be interpreted as rules of thumb without ceasing to be scientific are frozen and turned into measures of everything else – that is, when good science is turned into bad, because barren, ideology. (Unfortunately, many large-scale enterprises use this

21. An example is Needham's multi-volume study of Chinese science and technology; cf. especially his essay on Chinese medicine, *Celestial Lances*, Cambridge 1978. A summary with comments on the comparative authority of Western Science is *Science in Traditional China*, Cambridge, Mass. 1981. For 'primitive' cultures cf. Lévi-Strauss, *The Savage Mind*, and the vast literature that now exists in this field. Alexander Marshack, *Roots of Civilization*, New York 1972 and de Santillana and von Dechend, *Hamlet's Mill*, Boston 1969 discuss more recent approaches to palaeolithic art, technology and astronomy. For a survey and sociological considerations cf. C. Renfrew, *Before Civilization*, Cambridge 1979.

22. Thus Konrad Lorenz writes in his interesting and challenging if somewhat superficial book *Die Acht Todsünden der Zivilisierten Menschheit*, Piper 1984 (first published in 1973), p. 70: 'The erroneous belief that only what can be rationally grasped or even only what can be proved in a scientific way constitutes the solid knowledge of mankind has disastrous consequences. It prompts the "scientifically enlightened" younger generation to discard the immense treasures of knowledge and wisdom that are contained in the traditions of every ancient culture and in the teachings of the great world religions. Whoever thinks that all this is without significance, naturally succumbs to another, equally pernicious mistake, living in the conviction that science is able, as a matter of course, to create from nothing, and in a rational way, an entire culture with all its ingredients.' In a similar vein, J. Needham, *Time, The Refreshing River*, Nottingham 1986, speaks of 'Scientific Opium', meaning by it a 'blindness to the suffering of others'.

'Rationalism', writes Peter Medawar (*Advice to a Young Scientist*, New York 1979, p. 101), 'falls short of answering the many simple and childlike questions people like to ask: questions about origins and purposes such as are often contemptuously dismissed as non questions or pseudoquestions, although people understand them clearly enough and long to have an answer. These are intellectual pains that rationalists – like bad physicians confronted by ailments they cannot diagnose or cure – are apt to dismiss as "imagination" '. What I try to show in the present section is that questions concerning the value of science are exactly of this 'imaginary' kind.

Compare all this with E.O. Wilson's concession, quoted in footnote 2, that scientific materialism is lacking important dimensions of a comprehensive and also morally compelling world view. Note also that philosophers such as Kant tried to rescue the world-view character of science by showing how basic scientific principles are entrenched in human nature and therefore in life. (Popper, our own mini-Kant, and the proponents of an 'evolutionary epistemology', repeat the procedure on a more pedestrian level.)

ideology as one of their main intellectual weapons against opponents.) This concludes my discussion and defence of R1.

2 Political Consequences

Some industrial societies are democratic – they subject important matters to public debate – and pluralistic – they encourage the development of a variety of traditions. According to R1 each tradition may contribute to the welfare of individuals and of society as a whole. This suggests

R2: societies dedicated to freedom and democracy should be structured in a way that gives all traditions *equal opportunities*, i.e. equal access to federal funds, educational institutions, basic decisions. Science is to be treated as one tradition among many, not as a standard for judging what is and what is not, what can and what cannot be accepted.

Note that R2 is restricted to societies based on 'freedom and democracy'. This reflects my dislike of facile generalizations and my aversion to political actions based thereon. I do not favour the export of 'freedom' into regions that are doing well without it and whose inhabitants show no desire to change their ways. For me a declaration such as 'humanity is one, and he who cares for freedom and human rights cares for freedom and human rights everywhere', where 'to care' may imply active intervention (Christian Bay, *The Structure of Freedom*, New York 1968, p. 376), is just another example of intellectual (liberal) presumption. General ideas such as the idea of 'humanity', or the idea of 'freedom', or the Western idea of 'rights', arose in particular historical circumstances; their relevance for people with a different past must be checked by life, by extended contacts with their culture; it cannot be settled from afar. On the other hand, I think that the defenders of plurality, freedom and democracy have neglected some important implications of their creed. R2 describes the area of neglect and indicates how and on the basis of what arguments it can be reduced.

Note also that R2 recommends an equality of traditions and not only equality of access to one particular tradition ('equal opportunities' in Western democracies usually means the latter, the privileged tradition being a mixture of science, liberalism and capitalism). Traditions, not individuals are the units of discourse. To work, R2 must of course be made more specific. There have to be criteria for identifying traditions (not every association will count as a tradition and an entity that started as a tradition may deteriorate into a club) and regulating opportunities. But such criteria are better worked out by the groups that claim to be traditions and desire to have equal opportunities, rather than being stated in advance and independently of the parties concerned. The reason is that concrete political debates often lead to unforeseen changes of (a) the thoughts, customs, feelings that define the self image of a particular tradition, (b) the legal ideas (common law as well as statutory law) governing the treatment of traditions and (c) general (anthropological, historical, commonsense) ideas about the nature of traditions and cultures. The process needs lots of leeway to be acceptable to all participants. Political programmes and social theories that are conceived independently of it are too rigid to provide that leeway.

R2 demands equal opportunities and supports the demand by considering possible benefits: even the strangest ways of life *may have something to offer*. In the case of individuals, some authors have gone further, declaring that individuals *have rights* which do not depend on their usefulness. It seems natural to extend such rights to traditions, asserting, for example, that while we may learn a lot from the Mennonites, or the Shawnee, we should respect their ways even if they should turn out to be absolutely useless for the rest of society.[23] I therefore suggest that in addition to R1 and R2 we also postulate

> R3: Democratic societies should give all traditions *equal rights* and not only equal opportunities.

Again, the notion of rights and of an equality of rights must be further specified, and again the specification will have to come

23. Cf. Kant's request to treat (each part of) humanity as an end, never as a mere means: 'Act in such a manner that you regard humanity both in your own person and in the person of others always also as an end, and never merely as a means' (*Grundlegung der Metaphysik der Sitten*, 1786 (edition B), pp. 66f.). The request has famous ancestors. On 20 June 1500, the Catholic Crown in Spain formally approved liberty, not slavery, for Indians. Rafael Altamira commented as follows: 'What a memorable day for the entire world, because it signalises the first recognition of the respect due to the dignity and liberty of all men *no matter how primitive and uncivilised they may be* – a principle that had never been proclaimed before in any legislation, let alone practiced in any country' (quoted from Lewis Hanke, *All Mankind is One*, de Kalb 1974, p. 7).

from political debates between the concerned parties, from the proposal, discussion, critique of laws and precedents, not from armchair speculations. However some general consequences can be stated right away, and independently of more concrete developments.

For example, R3 implies that experts and governmental institutions in democratic societies must adapt their work to the traditions they serve instead of using institutional pressures to adapt the traditions to their work; medical institutions must take the religious taboos of special groups into account rather than trying to bring them in line with the most recent medical fashions. The suggestion is not at all unusual. Government scientists redefine their problems when a new administration comes along (or they are replaced by people with different convictions), scientists working on defence contracts adapt their approach to the changing political and defence climate, ecologists follow public needs, computer technologists switch their priorities with every flicker of the market, physiological research is prohibited by law from using live human beings or corpses not released for research by the relatives, and whoever has power imposes his idiosyncracies as if they were rights. R3 brings order into this practice and gives it what one might call a 'moral basis'. It also conforms with the decentralising tendencies inherent in every genuinely democratic society.

R2 and R3 are not absolute demands. They are proposals whose realization depends on the circumstances in which they are made and on the means (tactics, intelligence, power) available to those making them; they are open to modifications and exceptions. A defender of R3 differs from an opponent not by refusing to admit such exceptions, but by the fact that he treats them as exceptions, tries to do without them whenever possible, and keeps close to his ideal of equal opportunities and equal rights.[24]

R2 and R3 support both a *freedom for and a freedom from the sciences*: science, within our democracies, needs protection from non-scientific traditions (rationalism, Marxism, theological schools etc.) and non-scientific traditions need protection from

24. These remarks solve a problem created by the generalizing tendencies of some critics. 'If all traditions have equal rights', these critics say, 'then the tradition that only one tradition should have rights will have the same right as all others – which makes nonsense of R3!' But R3 is not a principle which 'entails' consequences; it is a rule of thumb which is made definite by its applications and which, apart from them, does not 'entail' anything.

science. Scientists may profit from a study of logic or of the Tao – but the study should emerge from scientific practice, it should not be imposed. Practitioners of traditional Chinese medicine may learn a lot from scientific approaches to human illness – but this learning process should again be permitted to proceed on its own, it should not be enforced by state institutions. Democratic decisions can of course impose (temporary) limits on any subject and any tradition; a free society, after all, must not be left at the mercy of the institutions it contains – it must supervise and control them. However such decisions come from debates in which none of the traditions plays a leading role (except accidentally and temporarily) and they can be reversed the moment they turn out to be impracticable or dangerous. I have discussed these and related problems in two books, *Against Method*, dealing with the freedom of science (from philosophical interference), and *Science in a Free Society*, dealing with the freedom of non-scientific traditions (from scientific interference).

3 Herodotus and Protagoras

Instead of discussing the exchange of customs, beliefs, ideas, we may ask how they affect people once they are *established* or, to use a slightly more abstract term, once they are believed to be *valid*. As far as I know the first writer to consider this was Herodotus. In book 3,38 of his *Histories* he tells the following story (quoted after the translation of Aubrey de Selincourt, Penguin Books 1954):

> When Darius was king of Persia, he summoned the Greeks who happened to be present at his court, and asked them what they would take to eat the dead bodies of their fathers. They replied they would not do it for any money in the world. Later, in the presence of the Greeks, and through an interpreter, so that they could understand what was said, he asked some Indians, of the tribe called Callatiae, who do in fact eat their parents' dead bodies, what they would take to burn them. They uttered a cry of horror and forbade him to mention such a dreadful thing. One can see by this what custom can do and Pindar, in my opinion, was right when he called it 'king of all'.

Custom is 'king of all' – but different people obey different kings:

> If anyone, no matter who, were given the opportunity of choosing from amongst all the nations of the world the set of beliefs which he

thought best, he would inevitably, after careful consideration of their relative merits, choose that of his own country. Everyone without exception believes his own native customs, and the religion he was brought up in, to be the best.

Now the rule of a king, generally, rests not only on power but also on rights. Herodotus implies that the same is true of customs. Invading Egypt, Cambyses tore down temples, ridiculed ancient laws, broke up tombs, examined the bodies and entered the temple of Hephaestus to mock the god's statue. Cambyses had the power to do all those things. Yet, according to Herodotus, he was not enlightened, he was 'completely mad; this is the only possible explanation of his assault upon, and mockery of, everything which ancient law and custom had made sacred in Egypt.' (Note that, following this line of reasoning, Xenophanes' mockery of 'everything which ancient law and custom have made sacred' also indicates a troubled and diseased mind; it is not a sign of enlightenment. Cf. the next chapter.)

To sum up:

> R4: laws, religious beliefs and customs rule, like kings, in restricted domains. Their rule rests on a twofold authority – on their *power* and on the fact that it is *rightful* power: the rules are *valid* in their domains.

R4 agrees with the views of Protagoras, Herodotus' great contemporary. In Plato's dialogue *Protagoras* the character 'Protagoras' explains his position twice; first, by telling a story, and then by 'giving reasons' (324d7). According to the story, Epimetheus was ordered by the gods and Prometheus to equip all creatures with suitable powers; not being a clever person, he exhausted the powers before he came to the human race which was left without protection and skills. To correct the mistake Prometheus stole fire and the arts from Hephaestus and Athena. Humans could now survive, but they were not yet able to live together in peace.

> Zeus, therefore, fearing the total destruction of our race, sent Hermes to impart to humans the quality of respect for others and a sense of justice and so to bring order into our cities, and create a bond of friendship and union.
>
> Hermes asked Zeus in what manner he was to bestow these gifts on the humans. 'Shall I distribute them as the arts were distributed – that is, on the principle that one trained doctor suffices for many laymen,

and so with the other experts? Shall I distribute justice and respect for their fellows in this way, or to all alike?'

'To all,' said Zeus. 'Let all have their share. There could never be cities if only a few shared in these virtues, as in the arts. Moreover, you must lay it down as my law that if anyone is incapable of acquiring his share of these two virtues he shall be put to death as a plague to the city.' (*Protagoras* 320dff, esp. 322cff, after Guthrie).

Justice, according to the story, is part of the law of Zeus. Laws and customs which are special articulations of justice thus again rest on a twofold authority: the power of human institutions, and the power of Zeus. 'Protagoras' explains how they are enforced:

As soon as a child can understand what is said to him, nurse, mother, tutor and the father himself vie with each other to make him as good as possible, instructing him through everything he does or says, pointing out 'this is right, that is wrong, this honorable and that disgraceful, this holy, that impious: do this, don't do that.' If he is obedient, well and good. If not, they straighten him with threats and beatings, like a warped and twisted plank (325c3ff).

As can be seen from these quotations Protagoras believed that there had to be laws and that they had to be enforced. He also believed that laws and institutions had to be adapted to the societies in which they were supposed to rule, that justice had to be defined 'relative to' the needs and the circumstances of these societies. Neither he nor Herodotus asserted, as did other sophists and later 'relativists', that institutions and laws that are valid in some societies and not valid in others are therefore arbitrary and can be changed at will. It is important to emphasize this point as many critics of relativism seem to take the inference for granted: one *can* be a relativist and yet defend and enforce laws and institutions.

The 'relativistic' ingredients of Protagoras' philosophy emerge from two sources. One is a report according to which Protagoras designed special laws for Thurii, a pan-Hellenic colony in Southern Italy. The second source is Plato's dialogue *Theaetetus* which contains a long discussion of ideas ascribed to Protagoras. The discussion starts with a statement which, being one of the few direct quotations we have from Protagoras, has become his trademark:

R5: Man is the measure of all things; of those that are that they are; and of those that are not, that they are not (152a1f).

As I pointed out in the last section and especially in footnote 24 (and text), a statement such as R5 can be interpreted in (at least) two ways: as a premise 'entailing' well defined and unambiguous consequences; or as a rule of thumb adumbrating an outlook without giving a precise description of it. In the first case (which is the one favoured by logicians) the meaning of the statement must be established *before* it is applied, or argued about; in the second case (which characterises most fruitful discussions, in the sciences and elsewhere) interpreting the statement *is part of* applying it, or arguing about it.[25] Plato leans towards the first interpretation, although the way in which R5 is being clarified (152a1-170a3) and the many asides that enliven the debate create the impression that he adopts the second. Yet the situation is clear enough: he wants a version of R5 that can be nailed down and refuted. The version he comes up with is

R5a: whatever seems to somebody, is to him to whom it seems (170a3f).

where 'whatever seems' means any (examined or unexamined) opinion that happens to appeal to the person in question.

Using this interpretation Plato produces three major criticisms of R5.

The first criticism starts with the observation (170a6ff) that few people trust their own opinions, that the great majority follows the advice of experts, that R5a, accordingly, is false for almost everybody and, as Protagoras measures truth by human opinions, also for him (171c5ff; cf. 179b6ff): 'he is caught when he ascribes truth to the opinions of others who give the lie direct to his own opinion.'

The second criticism is that experts produce reliable predictions while laymen do not (178a5ff). Thus in the case of medicine

25. The difference affects the translations. In the first case the translation must use precise terms; for example, it must make it clear if *anthropos* means a particular human being, or humanity in general, or an idealised thinking and judging being. In the second case a loose and open-ended translation will do the trick. The problems are described by Kurt von Fritz ('Protagoras', *Schriften zur Griechischen Logik* Vol. 1, Stuttgart 1978, pp. 111f.) as well as in Guthrie's philological comments, *A History of Greek Philosophy*, Vol. 3, Cambridge 1969, pp. 188ff. Charles H. Kahn, 'The Greek Verb "to be" and the Concept of Being', *Foundations of Language* Vol. 3 (1966), pp. 251f and *The Verb Be in Ancient Greek*, Dordrecht 1973, p. 376 has suggested, with plausible arguments, the translation 'Man measures what is so [is the case] that it is so [that it is the case] etc.'

'an ignoramus may believe that he will soon be suffering from a fever while a physician asserts the contrary; shall we say that the future will behave according to both opinions, or according to only one of them?' (178c2ff). Obviously the latter, says Plato – which removes R5a.

A third criticism deals with the structure of society. 'In social matters', says Socrates (1721ff), 'the theory [i.e. R5 read as R5a] will say that, as far as the Good and the Bad or the Just and Injust, or the Pious and the Impious are concerned, whatever opinion a state has in these matters and then lays down as the law, that also will be its Truth, and in these matters no individual or state is wiser than another.' But future developments may show the belief to be mistaken: some laws preserve the state while others blow it to pieces; some laws bring happiness to the citizens, while others create scarcity, strife, disaster. Truth, therefore, is not a matter of (individual, or collective, democratic, or aristocratic) opinion and R5a is false.

The arguments remove what R5a says about opinions (opinions are true for those who have them) by using opinions: that experts are better than laymen; that, being asked to solve a problem they will all give the same answer; that the answer will turn out to be correct in the future; that most people agree with this evaluation of experts – and so on.

But when Plato wrote his dialogue these special opinions had ceased to be popular and were attacked by experts as well as by laymen. For example, the author of the essay *Ancient Medicine* had ridiculed the tendency of medical theoreticians to replace commonsense by incomprehensible theories and to define sickness and health in their terms. A physician, said this writer, is supposed to restore the well being of people; he must therefore be able to state his aim in the same familiar words which are used by his patients (*Ancient Medicine*, chapters 15 and 20, quoted in section 6 below). Some physicians had argued and the public at large had come to believe that health returns to a sick organism all by itself and that experts are likely to retard the process (*On the Art*, chapter 5). Medicine, said the little treatise *Nomos* (chapter i) which seems to belong to the fourth century,

. . . is the most distinguished of all arts, but through the ignorance of those who practice it and of those who casually judge such practitioners, is now of all the arts by far the least esteemed. The chief reason for this error seems to me to be this: medicine is the only art which our states have made subject to no penalty save that of dishonor, and dishonor does not wound those who are full of it. Such men are in fact very like the supernumeraries in tragedies. Just as these have the appearance, dress and mask of an actor without being actors so, too, with the physicians; many are physicians by repute, very few are such in reality (after W.H.S. Jones, *Hippocrates*, Vol. ii, Loeb Classical Library, Cambridge 1967, p. 263).

Aristophanes (*Nub.*, 332f; cf. Victor Ehrenberg, *The People of Aristophanes*, New York 1962) classified 'quacks' together with prophets, playboys, 'dithyrambic pseudorythmicornament-producers' and 'shootingstarbegazing tricksters'. We learn from him that laymen often prepared their own medicine (*Thesmo-phoriazusae*, 483); and he wrote in his *Plutos* (407f):

Where can you find a doctor in this town?
The pay is low and low the outcome of their art.

This was also the time when old institutions were being gradually transformed and when increasing numbers of people participated in fundamental political decisions, decisions concerning the use of experts included. 'Our public men', said Pericles in his funeral oration (Thucydides, *Pelopponesian War*, 40, Modern Library College Edition),

have, besides politics, their private affairs to attend to, and our ordinary citizens, though occupied with the pursuits of industry, are still fair judges of public matters; for unlike any other nations, regarding him who takes no part in these duties not as unambitious but as useless, we Athenians are able to judge all events we cannot originate, and instead of looking at discussion as a stumbling block in the way of action, we think it an indispensable preliminary to any wise action at all. Again in our enterprises we present the singular spectacle of daring and deliberation, each carried to its highest point, and both united in the same persons . . . In short, I say that as a city we are the school of Hellas . . .

Trained in this school, Athenian citizens held opinions that had gone through a long process of adaptation, were well informed and very different from the isolated fits of 'seeming' Plato uses to interpret R5. Is it to be expected that Protagoras, who actively participated in the development, meant the latter when formulating his principle? Conversely, are not the special opinions Plato opposes to R5a precisely such fits and therefore suitable targets for his own criticism? Considerations such as these suggest that we read R5 in a less precise and technical manner and more in agreement with what I called above the second way of interpreting principles, for example as asserting

> R5b – that the laws, customs, facts that are being put before the citizens ultimately rest on the pronouncements, beliefs and perceptions of human beings and that important matters should therefore be referred to the (perceptions and thoughts of the) people concerned and not to abstract agencies and distant experts.

R5b is practical and realistic. It does not introduce artificial events and the situations mentioned in it depend on the problems that are being considered. For example, if we are talking about the laws of the city then the advice in the second part suggest that they should be 'measured' by the *citizens* rather than by gods or ancient lawgivers (who were traditionally regarded as their inventors and the sole source of their authority). If the 'important matters' are health and illness, then it is the *individual patient*, not a doctor engulfed in abstract theories, who 'measures' well being. 'Measuring' itself is no longer a process that compares a complex situation with the passing thoughts that happen to be in the minds of those involved in it; it includes learning (167blff). Some patients may of course cling to their inarticulate sense of well being, others may slavishly follow the prescriptions of their favourite doctors, but there will also be patients who read books, consult a variety of healers and come to a conclusion of their own. In politics some citizens may close their ears to argument and trust their 'gut feelings', others may listen to what the leading politicians have to say and then make up their minds. According to R5b all these individuals 'measure' what is being put before them.

Plato's three criticisms no longer apply to R5b.

The first criticism fails because 'opinion' now includes trust in the opinions of experts. The second criticism works in a community impressed by experts – which agrees with R5b. The third criticism collapses because future events, as interpreted by future generations, are just as much 'measures' as present events, interpreted by present observers: there exists no complete and stable knowledge of social and political matters. Summing up, we can say that Plato's major criticism rests on an unduly narrow interpretation of Protagoras' dictum combined with a dogmatic belief in the excellence of expert-knowledge.

My discussion of Plato's objections is not yet complete. Plato did not introduce experts because of the wonderful things they could do, but because he had an explanation for their success: experts know the truth and are in touch with reality. Truth and reality, not experts, are the ultimate measures of success and failure. Good ideas, procedures, laws, according to Plato, are neither popular ideas, procedures, laws, nor things that are supported by authorities such as kings, wandering bards, or experts; good ideas, procedures, laws are things that 'fit reality' and are true in this sense. Let us now see what R5 has to say about *this* point.

4 Truth and Reality in Protagoras

The distinction between being and seeming, truth and falsehood, the facts as they are and the facts as they are said or thought to be, was a familiar (though often only implied) part of common discourse long before Protagoras had formulated R5. 'As in the most contemporary idiom, so in Homer and Sophocles the man who speaks the truth "tells it like it is" and the liar tells it otherwise.'[26]

The presocratic philosophers, especially Parmenides, sharpened the distinction and made the duality (true-false) explicit. In addition they gave unified accounts of everything that could be said to be. These accounts conflicted with the unphilosophical ways of 'telling it like it is'. For the philosophers the conflict showed that commonsense was incapable of reaching truth. Democritus, for example, asserted that 'bitter and sweet are opinions, colour is an opinion – in truth there are atoms and the void',[27] while Parmenides rejected the 'ways of humans' (B1, 27), of 'the many' (B6, 7) who, being guided by 'habits based on much experience' (B7, 3), 'drift along, deaf as well as blind, disturbed and undecided' (B6, 6ff). Thus statements such as 'this is red', or 'that moves' which describe important events in the lives of artists, physicians, generals, navigators as well as ordinary human beings were summarily excluded from the domain of truth.

One of Protagoras' aims seems to have been to restore such

26. Kahn, *The Verb Be*, Dordrecht 1973, p. 363; cf. pp. 365, 369. For what follows see also Ch. 2 of Felix Heinimann, *Nomos and Physis*, Basel 1945 and Kurt von Fritz, 'Nous, Noein and their Derivatives in Presocratic Philosophy', *Classical Philology*, Vol. 40 (1945), pp. 223ff; Vol. 41 (1946), pp. 12ff.

statements to their former eminence. 'You and I,' Protagoras seems to say, 'our physicians, artists, artisans know many things and we live as we do because of this knowledge. Now these philosophers call our knowledge opinions based on shiftless experience and contrast "the many", i.e. people like us, with the enlightened few, i.e. themselves and their strange theories. Well, as far as I am concerned truth lies with us, with our "opinions" and "experiences" and we, "the many", not abstract theories, are the measure of things.'[28] Protagoras' reference to sensations (*Theaet.*, 152b1ff) can be seen in this light: 'sensations', for Protagoras, are neither the technical entities Plato constructs to get R5 into trouble (156a2ff) nor Ayerian sense-data; they are what common people rely on when judging their surroundings. Things are hot or cold for a person when the person feels them to be hot or cold, and not when a philosopher, using theory, pronounces the presence of The Hot or The Cold (two of Empedocles's abstract 'elements'). Protagoras's comments on mathematics (a circle cannot touch a ruler at only a single point – Aristotle, *Met.*, 998a) reflect the same attitude: practical concepts overrule concepts that have been separated from human action (modern constructivists proceed in an analogous way).[29] Both the arguments of the preceding section ('measuring' depends on circumstances; 'opinions' can be obtained in highly sophisticated ways) and the present considerations show that Protagoras reintroduced commonsense ways of establishing truth and defended them against the abstract claims of some of his predecessors. This, however, is not yet the whole story.

sense in matters of truth with rather uncommonsensical ideas

27. Diels-Kranz, Fragment B9. According to Reinhardt (V.E. Alfieri, *Atomos Idea*, Florence 1953, p. 127), the word *nomo* in the fragment is used in parallel to Parmenides' *nenomistai* (B6, 8) which in turn (Heinimann, *op. cit.*, 74ff) may be rendered as 'being customarily believed by the many' (but not true).

28. On the phrase 'the many' in Greek philosophical discourse from Homer to Aristotle cf. Hans-Dieter Voigtländer, *Der Philosophe und die Vielen*, Wiesbaden 1980. Protagoras is dealt with on pp. 81ff. 'Nobody can pretend', writes Victor Ehrenberg, *From Solon to Socrates*, Methuen, London and New York 1973, p. 340, 'that the sentence [R5] or its translations is clear and meaningful. It needs further explanation and that is by no means obvious . . . It is likely that Protagoras went beyond the meaning of mere sense perception . . . The main point, the one clearly positive and the one which impressed people at once and for all time, is the *metron anthropos*, the central position given to man.' I would add: to man insofar as he is engaged in his ordinary, day-to-day activities, not to man the inventor of abstract theories.

about falsehood. According to *Euthydemus* 286c and *Theaet.* 167a7ff., he thought it impossible to (try to speak truthfully and yet) make a false assertion. It seems that this doctrine was connected with the idea, found in Parmenides (B 2,8; B 8,7) and exploited by Gorgias (*On the Non Existent or on Nature*), that false statements, being about nothing, also say nothing: perception and opinion, the customary measures of truth, are infallible measures and the worlds projected by different individuals, groups, nations are as they perceive and describe them – they are all equally real. However, *they are not equally good or beneficial* (to those who live in them). A sick person lives in a world where everything tastes sour and therefore is sour (166e2ff.) – but he is not happy in it. The members of a racist society live in a world where people fall into sharply defined groups, some creative and benevolent, others parasitical and evil – but their lives are not very comfortable. A desire for change may arise in either case. How can the change be effected?

According to Protagoras changes are caused by wise men (166d1ff.). Wise men cannot change falsehood into truth or appearance into reality – but they can change an uncomfortable, painful and threatening reality into a better world. Just as a physician, using medicine, changes a real but distressful state of an individual into an equally real but agreeable state (of the same individual – or of a changed individual), in the same way a wise man, using words, changes an evil and ruinous state (of an

29. This interpretation was suggested by E. Kapp, *Gnomon*, Vol. 12 (1936), pp. 70ff. Kurt von Fritz adopts Kapp's views (art. cit., fn. 144); in his article 'Protagoras' (cf. fn. 25) he compares Protagoras' statement with the complaints of the author of *Ancient Medicine* that medical theoreticians describe illnesses and cures in terms of abstract entities such as The Hot, The Cold, The Wet, The Dry without saying a word about the particular food (hot milk? lukewarm water?) that is supposed to be taken or the particular ailment (diarrhoea) that affects the patient. Considering such parallels, von Fritz infers (p, 114) that Protagoras' statement 'was not originally designed to formulate a consistent sensualism, relativism or subjectivism but rather wanted to confront the strange philosophy of the Eleatics [according to whom Being had not part and did not change], or Heraclitus [according to whom there was only change] and of others who had left the communis opinio far behind with a commonsense philosophy just as the author of [*Ancient Medicine*] confronted a medical school which derived its science from general philosophical and scientific principles with a purely empirical medicine and added explicitly that a medical theory could be of value only if comprehensible for a layman.' Cf. also F.M. Cornford, *Plato's Theory of Knowledge*, New York 1957, p. 69: 'All that the objections [raised in *Theaet.* 164c–165e] in fact established was that "perception" must be stretched to include awareness of memory images.'

individual or of an entire city) into a beneficial state. Note that according to this account it is the individual or the state, not the wise man, who judges the success of the procedure. Note also that this judgement, acting back on the wise man, may improve his own state of expertise and thus turn him into a better advisor. Note, finally, that in a democracy the 'wise man' is the community of citizens as represented by the general assembly: what the assembly says is both a truth about society and an instrument for changing it, and the reality brought about by its statements is an instrument for changing the procedures and the opinions of the assembly in turn. This is how Protagoras' theory of truth and reality can be used to explain the workings of a direct democracy.

It is interesting to compare Protagoras's views with the more familiar forms of philosophical and scientific objectivism. Objectivism asserts that everybody, no matter what their perceptions and opinions, lives in the same world. Special groups (astronomers, physicists, chemists, biologists) explore this world, other special groups (politicians, industrialists, religious leaders) make sure that people can survive in it. First the manufacturers of an objective reality go on their flights of fancy, then the material and social engineers connect the results with the needs and wishes of the common folk, that is, with reality as defined by Protagoras. Protagoras collapses the two procedures into one: 'reality' (to speak the objectivists' language) is explored by attempting to satisfy human wishes in a more direct way; thought and emotions work together (and are perhaps not even separated). We might say that the approach of Protagoras is an engineering approach, while the objectivists who separate theory and practice, thought and emotion, nature and society, and who carefully distinguish between objective reality on the one side and experience and everyday life on the other, introduce sizeable metaphysical components. Trying to change their surroundings so that they look more and more like this reality (and thus make them feel comfortable), the objectivists act of course like pure Protagoreans, but not like Protagorean wise men. To become wise, they must 'relativize' their approach. There are many indications that this is already part of their practice.

To start with, objectivists have not constructed one world, but many. Of course, some of these worlds are more popular than others, but this is due to a preference for certain values (in addition to the value of objectivity – see section 2), not to intrinsic advantages: results of measurement are preferred to qualities because technological changes are preferred to harmonious adaptations; laws of nature overrule divine principles because they act in a more monotonous way – and so on. The plurality affects the sciences which contain highly valued experimental enterprises such as molecular biology side by side with despised qualitative disciplines such as botany or rheology. The most fundamental science, physics, has so far failed to give us a unified account of space, time and matter. What we have, therefore (apart from grandiloquent promises and superficial popularizations), is a variety of approaches based on a variety of models and successful in restricted domains, i.e. what we have is a Protagorean practice.

Secondly, the transition from a particular model to practical matters often involves such sizeable modifications that we would do better to speak of an entirely new world. Industry in various countries confirms this conjecture by decoupling its research from universities and engineering schools and developing procedures more suited to its own particular needs. Social programmes, ecological studies, impact reports for technological projects often raise problems unanswered by any existing science; those engaged in the studies are forced to extrapolate, redraw boundaries, or develop entirely new ideas to overcome the limits of specialist knowledge. Thirdly, objectivist approaches, especially in health, agriculture and social engineering, may succeed by forcing reality into their patterns; the distorted societies then start showing traces of the patterns imposed. This is again a truly Protagorean procedure except that it inverts the Protagorean chain of command: what counts is the judgement of the interfering scientists and not the judgement of the people interfered with. Fourth, the interference often upsets a delicate equilibrium of aims and means and so does more harm than good – and this is now recognized by the 'developers' themselves. In his study *The Constitution of Liberty* Chicago 1960, chapter 4, p. 54, F.A. von Hayek distinguishes between what he calls 'two different traditions in the theory of liberty', 'one empirical and unsystematic, the other speculative and rationalistic – the first based on an interpretation of traditions and institutions which had spontaneously grown up and were but imperfectly understood, the second aiming at the construction of a Utopia, which has often been tried, but never successfully', and he explains why the former is to be preferred to the latter. But the former tradition is closely related to the Protagorean point of view whose 'seeming' reflects partly comprehended, partly unnoticed adaptations to what the nature of the moment happens to be. If debates play an important role in the adaptations and if the debates are carried out by an assembly of free citizens so that everybody has the right to act as a 'wise man', then we have what I shall call a *democratic relativism*. In the next section I shall describe this form of society in somewhat greater detail. Before that, however, I wish to make some comments on the notion of a debate.

One of the main objections against Protagoras is that different Protagorean worlds cannot clash and that debates between their inhabitants are therefore impossible. This may be true for an outside observer; it is not true for the participants who, perceiving a conflict, can start a quarrel without asking his permission. The parties to a debate (I shall call them A and B) need not share any elements (meanings, intentions, propositions) that can be detached from their interaction and examined independently of the role they play in it. Even if such elements existed the question would still arise how, being outside human lives, they can enter them and affect them in the specific way in which an assertion, or a thesis, or a belief affects the consciousness and the actions of the participants. What is needed is that A has the impression that B shares something with him, seems to be aware of this, acts accordingly; that a semanticist C, examining A and B, can develop a theory of what is shared and how what is shared affects the conversation; and that A and B, on reading C, have the impression that he has hit the nail on the head. Actually, what is needed is much less: A and B need not accept what they find when reading C – C may still survive and be respected if there exists a profession that values his ideas. Reputations, after all, are made and broken by the impression the actions of some people make on others. Appeals to a higher authority are empty words unless the authority is noticed, i.e. appears in the consciousness of the one or the other individual.

5   Democratic Relativism

R5, interpreted as R5b, is of far greater importance than modern philosophical analyses and 'clarifications' would make us believe. It can guide people in their dealings with nature, social institutions and with each other. To explain I shall first give some historical background.

Most societies that depend on a close collaboration between

diverse groups have experts, people with special knowledge and special skills. Hunters and gatherers, it seems, possessed all the knowledge and all the skills necessary for survival. Large scale hunts and agriculture then led to a division of labour and tighter social controls. Experts arose from this development: the Homeric warriors were experts in the conduct of war; rulers like Agamemnon, in addition, knew how to unite different tribes under a single purpose; physicians healed bodies, mantics interpreted omens and predicted the future. The social position of experts did not always correspond to the importance of their services. Warriors might be servants of society, to be called upon in times of danger, but without special powers in times of peace; on the other hand, they might be its masters, shaping it in accordance with their own warlike ideology. Scientists once had no greater influence than plumbers; today large sections of society reflect their view of things. Experts were a matter of course in Egypt, Sumeria, Babylonia, Assyria, among the Hittites, the Hurrites, the Phoenicians and the many other peoples that populated the ancient Near East. They played an important role in the Stone Age as is shown by the amazing remnants of Stone Age astronomy and Stone Age mathematics that have been discovered over the years. The first recorded discussion of the problems of expert knowledge occurred in Greece, in the fifth and fourth centuries B.C., among the sophists, and then in Plato and Aristotle.

The discussion anticipated most modern problems and positions. The ideas it produced are simple and straightforward and unencumbered by the useless technicalities of modern intellectual debates. We can all learn from these old thinkers, their argument and their views.

The discussion also went beyond the authority of special fields, such as medicine and navigation; it included inquiries concerning the good life and the right form of government: should a city be governed by a traditional authority such as a king, or by a board of political experts, or should government be a matter for all?

Two views emerged from the discussion. According to the *first view* an expert is a person who produces important knowledge and has important skills. His knowledge and his skills must not be questioned or changed by non-experts. They must be taken over by society in precisely the form suggested by the experts. High priests, kings, architects, physicians occasionally saw their function in this way – and so did some of the societies in which they worked. In Greece (Athens, fifth century B.C.) this view was an object of ridicule.[30]

Representatives of the *second view* pointed out that experts in arriving at their results often restrict their vision. They do not study all phenomena but only those in a special field; and they do not examine all aspects of these special phenomena but only those related to their occasionally rather narrow interests. It would therefore be foolish to regard expert ideas as 'true', or as 'real' – period – without further studies that go beyond expert limits. And it would be equally foolish to introduce them into society without having made sure that the professional aims of the experts agree with the aims of society. Even politicians cannot be left unattended, for though they deal with society as a whole they deal with it in a narrow way, being guided by party interests and superstitions and only rarely by what others might regard as 'true knowledge'.

According to Plato, who held the view just described, the further studies are the task of super-experts, namely philosophers. Philosophers define what it means to know and what is good for society. Many intellectuals favour this *authoritarian approach*. They may overflow with concern for their fellow human beings, they may speak of 'truth', 'reason', 'objectivity', even of 'liberty', but what they really want is the power to reshape the world in their own image. There is no reason to assume that this image will be less one-sided than the ideas it wants to control and so it, too, must be examined. But who will carry out *this* examination? And how can we be sure that the authority to which we entrust the matter does not again introduce its own narrow conceptions?

The answer given by the *democratic approach* (in a sense to be clarified as the argument proceeds) arose in particular historical circumstances. 'Natural' societies 'grew' without much conscious planning on the part of those who lived in them. In Greece major changes, in special fields as well as in society at large, gradually became a matter of debate and explicit reconstruction. Athenian democracy at the time of Pericles took care that every free man could have his say in the debate and could temporarily assume any position, however powerful. We do not know the steps that led to this very specific type of adaptation and it is by no means certain that the development was beneficial in all respects. Some

30. Cf. e.g. J. Burkhardt, *Griechische Kulturgeschichte*, Munich 1977, Vol. 4, pp. 118ff.

of the difficulties that trouble us today suggest that debate and 'rational discourse' in particular are not a universal panacaea, that they may be too crude to capture the more subtle threats to our well being and that there may be better ways of conducting the business of life.[31] But societies that are committed to it and define liberty and a worthwhile life accordingly, cannot exclude a single opinion, however outlandish. For what are political debates about? They are about the needs and the wishes of the citizens. And who are better judges of these needs and wishes than the citizens themselves? It is absurd first to declare that a society serves the needs of 'the people' and then to let autistic experts (liberals, Marxists, Freudians, sociologists of all persuasions) decide what 'the people' 'really' need and want. Of course, popular wishes have to take the world into account and this means: the available resources, the intentions of the neighbours, their weapons, their policies – even the possibility that strong popular desires and aversions are unconscious and accessible to special methods only. According to Plato and his modern successors (scientists, politicians, business leaders), it is here that the need for expert advice arises. But experts are just as confused about fundamental matters as those they are supposed to advise and the variety of their suggestions is at least as large as the variety implicit in public opinion.[32] They often commit grievous mistakes. Besides, they never consider all aspects that affect the rest of the population but only those that happen to correspond to the current state of their speciality. This state is often far removed from the problems faced by the citizens. Citizens, guided but not replaced by experts, can pinpoint such short-comings and work towards their removal.[33] Every trial by jury provides examples of the limits and contradictions inherent in

31. The problem alluded to forms a vast subject which we are only slowly beginning to understand. For example, it is becoming clear that the difficulties of some so-called 'Third World' countries may be results of the manipulative rationality of the West rather than of an original barrenness of the land, or of the incompetence of those tending it. The expansion of Western civilization robbed many indigenous people of their dignity and their means of survival. Wars, slavery, simple murder was for a long time the right way of dealing with 'primitives'. But the humanitarians have not always fared better than the gangsters. Imposing their own views of what it means to be human and what a good life consists in, they have often added to the destruction wrought by their colonial predecessors. For details, see the literature in footnote 9.

32. As an example consider the many ways in which Freudians, existentialists, geneticists, behaviourists, neurophysiologists, Marxists, theologians (hardcore Catholics; liberation theologists) define human nature and the great variety of suggestions they make on topics such as education, war, crime etc.

expert testimony and encourages the jurors to make reasonable guesses in recondite fields. The citizens of a democracy, Protagoras would say, expressing the political ideas of Periclean Athens (which differed from the science-ridden democracies of today and was less inhibited by restrictions), receive this kind of education not merely once or twice in their lifetime, but every single day of their lives. They live in a state – small and manageable Athens – where information freely flows from one citizen to the next. They not only *live* in this state, they also *conduct its business*; they discuss important problems in the general assembly and occasionally lead the discussions, they participate in law courts and artistic competitions, they judge the work of writers who are now regarded as some of the greatest dramatists of 'civilised humanity' (Aeschylus, Sophocles, Euripides, Aristophanes all competed for public prizes), they initiate and terminate wars and auxiliary expeditions, they receive and examine the reports of generals, navigators, architects, food merchants, they arrange foreign aid, welcome foreign dignitaries, listen to and debate with sophists, garrulous Socrates included – and so on. They use experts all the time – but in an advisory capacity, and they make the final decisions themselves. According to Protagoras the knowledge the citizens acquire during this unstructured but rich, complex and active process of learning (learning is not separated from living, it is part of it – the citizens learn while carrying out the duties that need the knowledge acquired) suffices for judging all events in the city, the most complex technical problems included. Examining a particular situation (such as the danger of a meltdown in a nearby nuclear reactor – to use a modern example), the citizens will, of course, have to study new things – but they have acquired a facility of picking up unusual items and, most importantly, they have a sense of perspective that allows them to see the strong points and the limitations of the proposals under review. No doubt the citizens will commit mistakes – everybody commits mistakes – and they will suffer from them. But suffering from their mistakes they will also become wiser while the mistakes of experts, being

33. Robert Jungk, in an interesting and provocative book about nuclear power (*The New Tyranny*, New York 1979), reports that citizens are often better informed about relevant scientific literature than scientists and that, having different and wider interests (for example, they are interested in the future well being of their children), they may consider effects not yet examined by scientists. A concrete example of the impact of citizens' initiatives is examined in R. Meehan, *The Atom and the Fault*, Cambridge, Mass. 1984.

hidden away, create trouble for everybody but enlighten only a privileged few. We may sum up this point of view by declaring that

> R6: citizens, and not special groups have the last word in deciding what is true or false, useful or useless for their society.

So far, a short and very sketchy account of ideas that are found, in traces, in Protagoras and in Periclean Athens. I shall call the point of view they adumbrate *democratic relativism*.

Democratic relativism is a form of *relativism*; it says that different cities (different societies) may look at the world in different ways and regard different things as acceptable. It is *democratic* because basic assumptions are (in principle) debated and decided upon by all citizens. Democratic relativism has much to recommend it, especially for us in the West, but it is not the one and only possible way of living. Many societies are built up in a different way and yet provide a home and means of survival for their inhabitants (see the comments on R2 as well as footnote 31 and text).

Democratic relativism has interesting ancestors, the *Oresteia* of Aeschylus among them: Orestes avenged his father; this satisfies the law of Zeus, represented by Apollo. To avenge his father Orestes has to kill his mother; this mobilises the Eumenides who oppose the murder of blood relatives. Orestes flees and seeks protection at the altar of Athene. To solve the problem created by the conflicting moralities, Athene initiates a 'rational debate' between Apollo and the Eumenides, with Orestes participating. Part of the debate is the question of whether a mother is a blood relative. The Eumenides say she is: Orestes spilt the blood of a blood relative and must be punished. Apollo says she is not: the mother provides warmth, protection and nourishment for the seed, she is a breeding oven, but she does not contribute her blood to the child (this view was held for a long time after). Today the debate would be resolved by experiments and expert judgement: experts would withdraw into their laboratories and Apollo, Orestes and Athene would have to wait for their findings. In Aeschylus the matter is decided by a vote: a court of Athenian citizens is informed about the case and gives its opinion. The votes are equally balanced after Athene adds her own vote in favour of Orestes (she was born without a mother) and so Orestes is released from the revenge of the Eumenides. But Athene also declares that their world view will not be dis-

carded: the city needs *all* the agencies that made it grow, it cannot afford to lose a single one of them. True, there are now new laws and a new morality – the laws of Zeus as represented by Apollo. But these laws are not permitted to sweep aside what came before. They are granted entrance into the city *provided* they share the power with their predecessors. Thus a generation before Herodotus popular laws and customs were declared to be *valid* while their validity was *restricted* to make room for other, but equally important laws and customs. (Note also the similarity with Mill's philosophy as described in footnote 16 and text.)

Democratic relativism does not exclude the search for an objective, i.e. a thought-, perception- and society-independent reality. It welcomes research dedicated to finding objective facts but controls it by (subjective) public opinion. It thus denies that showing the objectivity of a result means showing that it is binding for all. Objectivism is treated as one tradition among many, not as a basic structure of society. There is no reason to be troubled by such a procedure and to fear that it will destroy important achievements. For although objectivists have discovered, delineated and presented situations and facts that exist and develop independently of the act of discovery, they cannot guarantee that the situations and facts are also independent of the entire tradition that led to their discovery (cf. section 9). Besides, even the most determined (and best paid) application of what many Western intellectuals regard as the most advanced versions of objective research have so far failed to give us the unity the idea of a universal and objective truth suggests. There are grandiloquent promises, there are blunt assertions of unifications already achieved but what we have *in fact* are regions of knowledge similar in structure to the regionalism Herodotus described so vividly in his history. Physics, the alleged core of chemistry and, via chemistry, of biology, has at least three principal subdivisions: the domain of the very large ruled by gravitation and tamed by Einstein's theory of general relativity (and various modifications thereof); the domain of the very small, ruled by the strong nuclear forces but not yet tamed by any comprehensive theory (the 'Grand Unified Theories' or 'GUTs' are, according to Gell-Mann, 'neither grand, nor unified; it might even be said that they are not even theories – just glorified models'); and, finally, an intermediate domain where quantum theory reigns supreme. Outside physics we have qualitative knowledge which contains commonsense and parts of biology, chemistry, geology as yet unreduced to the 'basic science' of the moment. The theories or points of view that define the processes in all these domains either clash, or cease to make sense when universalised, i.e. when assumed to be valid under all circumstances. Hence we may either interpret them as instruments of prediction with no relevance for what is true, or real, or we may say that they are 'true for' special areas which are defined by special questions, procedures, principles. Alternatively, we may assert that one theory reflects the basic structure of the world while the others deal with secondary phenomena. In this case speculation rather than empirical research becomes the measure of truth. Pluralism survives, but it is lifted on to the metaphysical plane. Speaking in the manner of Herodotus we can summarise the situation in the following way:

R7: the world, as described by our scientists and anthropologists, consists of (social and physical) regions with specific laws and conceptions of reality. In the social domain we have relatively stable societies which have demonstrated an ability to survive in their own particular surroundings and possess great adaptive powers. In the physical domain we have different points of view, valid in different areas, but inapplicable outside. Some of these points of view are more detailed – these are our scientific theories; others simpler, but more general – these are the various philosophical or commonsense views that affect the construction of 'reality'. The attempt to enforce a universal truth (a universal way of finding truth) has led to disasters in the social domain and to empty formalisms combined with never-to-be-fulfilled promises in the natural sciences.

Note that *R7 is not meant to be read as a universal truth*. It is a statement made within a particular tradition (Western intellectual debate starting from and leading towards scientific results), explained and defended (more or less competently) according to the rules of this tradition and indicating that the tradition is incoherent. The statement is of no interest to a Pygmy, or to a follower of Lao-Tsu (although the latter may study it for historical reasons). Note also that parts of R7 depend on a special evaluation of knowledge claims: quantum mechanics and relativity are assumed to offer equally important, equally successful, and equally acceptable accounts of the material universe. Some critics (Einstein among them) judge the situation differently. For them relativity physics goes to the bottom of things while the quantum theory is an important but highly unsatisfactory prelude to more substantial views. These physicists reject R7 and assert that universally valid theories already exist. As I said above this introduces metaphysical conjectures where assertions concerning objectivity depend on a subjective weighing of knowledge claims. There are again many such approaches (the orthodox one among them) which means that plurality is transformed (it is made metaphysical), it is not removed. In the next section I shall comment on this feature of the debate.

Democratic relativism is not the philosophy that guides modern 'democracies': power, here, is delegated to distant power centres, and important decisions are made by experts, or the 'representatives of the people', hardly ever by 'the people' themselves. Still, it seems a good starting point for Western intellectuals trying to improve their own life and the lives of their fellow human beings (it seems a good starting point for citizens' initiatives). It encourages debate, argument, and social reconstruction based on both. It is a specific political view, restricted in appeal and not necessarily better than the more intuitive procedures of 'primitive' societies. Yet since it invites the participation of all it may lead to the discovery that there are many ways of being in the world, that people have a right to use the ways that appeal to them and that using these ways they may lead a happy and fulfilling life.[34]

6   Truth and Reality: Historical Treatment

In the preceding sections the nature of the interaction between cultures was left unspecified. For example, no conditions were imposed on the studies and the possible gains mentioned in R1. In section 5 this was shown to be an essential part of the democratic approach: if the members of a tribe, a culture or a civiliz-

---

34. Some modern liberals grant foreign cultures the right to exist provided they participate in international trade, permit Western doctors to heal them and Western missionaries (of science and of other religions) to explain the wonders of science and Christianity to their children. But the idea of a peaceful Commonwealth of Nations whose members learn from each other thus constantly rising to new stages of knowledge and awareness is not shared by the Pygmies (for example) who prefer to be left alone (C.M. Turnbull, 'The Lesson of the Pygmies', *Scientific American* 208 (1), 1963). Rationalists such as Karl Popper (*The Open Society and its Enemies*, Vol. 1., New York 1963, p. 118) have no objection to applying pressure at this point: the entrance into mature humanity may have to be enforced 'by some form of imperialism'. I don't think that the achievements of science and rationalism are sufficiently dazzling to justify such a procedure.

ation have the impression that they have profited from the exchange and that their lives have improved, then this already settles the matter; cultural exchange is the business of the participants, not of outsiders (except when the exchange prepares a war against them).

Many intellectuals disagree. They warn us that what may seem like tremendous advantages to those who are engaged in the exchange and who conduct it according to their lights may in fact be grievous mistakes.

As it stands the warning is hardly needed. There exists no society that does not have a notion of error and procedures for discovering and rectifying mistakes. But the intellectuals who issue the warning define errors in a special way; they define them not by reference to the (standards and procedures of the) form of life in which they occur but by comparison with a society-independent 'reality', 'rationality', or 'truth'. Using these measures they have condemned entire cultures as being based on illusion and prejudice. The philosophical (as opposed to the practical) versions of relativism try to block such moves either by offering relativistic analyses of truth, reality and rationality or by devising alternative notions. Needless to say, they are rather complicated. In section 4 I discussed some ancient approaches to the problem. I now add some historical comments.

Like many other notions that were appropriated and transformed by spiritual leaders (prophets, scientists, philosophers, run-of-the-mill intellectuals), the ideas of truth, reality and rationality make excellent practical sense.

For example, telling the truth usually means telling what happened in a particular situation; it means 'telling it like it is' (cf. section 4). The person asked may not have the necessary information – the reply will then be 'I do not know' or 'I cannot really say'. But there are cases where the witness can give an answer and would be rightly called a liar if he said that he did not know. These cases may be doubted in turn: the identified individual may be an identical twin, the witness may have been looking at a mirror and not at a real person and so on. Yet the request to 'speak the truth' makes sense just as it makes sense to speak of real things despite all the illusions an inventive magician might conjure up.

For example, it makes sense to say that the room in which I am now sitting is real but that the room in which yesterday, in a dream, I saw an elephant riding on a sparrow was not. Of course, a dream is not a nothing; it may have important consequences for the dreamer and for others (the dreams of kings decided about war and peace, life and death). But the effects of a dreamed event upon the waking world differ from the effects of a perceived event; some cultures express the difference by saying that the dreamed event is not 'real'.

The notion of reality which underlies this way of drawing lines cannot be explained in a simple definition. A rainbow seems to be a perfectly real phenomenon. It can be seen, it can be painted, it can be photographed. However, we cannot run into it. This suggests that it is not like a table. It is not like a cloud either, for a cloud does not change position with the movement of the observer as a rainbow does. The discovery that a rainbow is caused by light being refracted and reflected inside water droplets reintroduces clouds together with an explanation of the peculiarities of rainbows thus returning to them at least part of the reality of clouds: grand subdivisions such as the subdivision real/unreal are much too simplistic to capture the complexities of our world. There are many different types of events and 'reality' is best attributed to an event together with a type, and not absolutely. But this means that we need only types and their relations and can dispense with 'reality' altogether.

Commonsense views (tribal commonsense; the use of common notions in modern languages) are built in precisely this manner. They contain subtly articulated ontologies including spirits, dreams, battles, ideas, gods, rainbows, pains, minerals, planets, animals, festivities, justice, fate, sickness, divorces, the sky, death, fear – and so on. Each entity behaves in a complex and characteristic way which, though conforming to a pattern, constantly reveals new and surprising features and thus cannot be captured in a formula; it affects, and is affected by, other entities and processes constituting a rich and varied universe. In such a universe the problem is not what is 'real' and what is not: queries like these do not even count as genuine questions. The problem is what occurs, in what connection, who was, or could be misled by the event and how.

'Problems of reality' arise when the ingredients of complex worlds of this kind are subsumed under abstract concepts and are then evaluated, i.e. declared to be either 'real' or 'unreal' on that basis. They are not fruits of more refined ways of thinking; they arise because delicate matters are compared with crude ideas and are found to be lacking in crudeness.

We can occasionally explain why crude ideas get the upper hand: special groups want to create a new tribal identity or preserve an existing identity amidst a rich and varied cultural landscape; to do so they excise large parts of the landscape and either disregard their existence or make them wholly evil. The first procedure was chosen by the Israelites at the time of Moses (monotheism), the second by the early Christians. For some Gnostics the entire 'material world' (in itself a gross simplification) was evil deception. Crude ideas may lead to limited successes; this encourages the proponents and reinforces their ways of thinking (consider, for example, the enthusiasm for quantification and the contempt for qualitative considerations among many scientists): ontological sophistication is a luxury when the survival of a tribe or a religious group, or the reputation of a well paid profession, is at stake.[35]

The 'rise of rationalism' in ancient Greece is a fascinating example of this attempt to transcend, devalue, and push aside complex forms of thought and experience. Having some details

35. In his book *The Religion of Israel*, New York 1972, Yehezkel Kaufmann describes the cultural environment of Israel (chapter 6) and comments on the fact that 'although the whole of biblical literature is a product of the deep transformation [Moses's message] brought about, the Bible tells nothing of the course of that transformation' (p. 230). 'The Bible's ignorance of the meaning of paganism', writes Kaufmann (p. 20), 'is . . . the basic problem . . . to the understanding of biblical religion.' However it is also its 'most important clue' (*ibid.*), indicating that the change had not just led to a loss of power of the pagan gods but to their complete disappearance: 'The Bible nowhere denies the existence of the gods; it ignores them. In contrast to the philosophic attack on Greek popular religion and in contrast to the later Jewish and Christian polemics, biblical religion shows no trace of having undertaken deliberately to suppress and repudiate mythology' (p. 20).

St. Paul, on the other hand, declared the pagan gods to be demons: I *Cor.* x, 20.

Gnosticism is a complex matter and gnostic theories and myths are among the most colourful productions of the human mind. Yet grand chasms subdivide the world in all its versions: cf. R.M. Grant, *Gnosticism and Early Christianity*, New York 1966.

Reducing phenomena to a few principles succeeded in parts of physics and in the astronomy of the planetary system. The extension to medicine was a disaster. 'In addition to their feeble logic' writes R.H. Shryock (*The Development of Modern Medicine*, The University of Wisconsin Press 1936, p. 31), commenting on physicians who tried to reform medicine in analogy with Newton's unification of planetary astronomy, 'systematicists exhibited characteristic personal failings. These were egotism, pride in a system as a sort of artistic creation, zeal to establish it as one spreads a new gospel, and a notable desire to defend it against all comers. Such straits accounted for the bitter professional controversies in which some physicians indulged. Obviously, if one philosopher was right, the others were all wrong, and in such cases it was difficult to restrain one's feelings. Dogmatism in medicine was no more inclined to toleration than was dogmatism in theology . . . '

at our disposal we notice that it was not a simple process but involved different strands which, being amplified by a kind of resonance, led to major historical changes. The most obvious intellectual manifestations of the change were the opinions of writers such as Anaximander, Heraclitos, Xenophanes and Parmenides. These writers affected history not by the power of their ideas but because of concurrent tendencies towards generalization and abstraction. Without any help from the philosophers, 'words . . . [had] become impoverished in content, they had become one sided and empty formulae'.[36] The deterioration was noticeable in Homer; it became prominent in Hesiod, and it was obvious in the Ionian philosophers of nature, in historians such as Hekataeus and in certain passages of the (epic, tragic, lyrical, comic) poets. In politics abstract groups replaced neighbourhoods as the units of political action (Cleisthenes), in economics money succeeded barter, the relations between military leaders and soldiers became increasingly impersonal and uniform, life as a whole moved away from personal relations and terms involving such relations either lost in content or disappeared. Small wonder the extreme views of the early philosophers found followers and could start a trend.

Further help came from the discovery (which seems to have occurred some time between Xenophanes and Parmenides) that statements composed of concepts lacking in details could be used to build new kinds of stories, soon to be called proofs, whose truth 'followed from' their inner structure and needed no support from traditional authorities. The discovery was interpreted as showing that knowledge could be detached from traditions and made 'objective'. Cultural variety, I said in the introduction, generates a variety of reactions, from fear and aversion to curiosity and the wish to learn, and a corresponding variety of doctrines ranging from extremely xenophobic forms of dogmatism to equally extreme forms of relativism and opportunism. The existence of proofs (or of weaker, but equally 'rational' forms of argumentation) apparently put an end to this confusion; it seemed that all one had to do was to accept what had been proved and reject the rest – and truth would

36. Kurt von Fritz, *Philosophie und Sprachlicher Ausdruck bei Demokrit, Platon und Aristoteles*, Neudruck Darmstadt 1966, 11. For what follows cf. also Bruno Snell *The Discovery of the Mind*, New York 1960, passim, together with the enlarged fourth edition *Die Entdeckung des Geistes*, Göttingen 1975. For explanations and parallel trends see E.G. Forrest, *The Emergence of Greek Democracy*, London 1966.

emerge in a culture-independent way.

The leading representative of this view was Parmenides. In the poem in which he explains his ideas he distinguishes between two procedures or 'pathways', as he calls them. The one, based on 'habit, born of much experience' (*ethos polypeiron*: B 7,3), i.e. on traditional forms of knowledge and knowledge-acquisition, contains the 'opinions of mortals'; the other, 'far from the foot-steps of humans' (i.e. independent of traditions), leads to what is 'appropriate and necessary'. According to Parmenides the second path is not a tradition, but supersedes all traditions.[37] Many scientists seem to view their activity in a similar way.

The view is clearly mistaken.

We may agree that abstract notions and principles can be connected more easily than practical (empirical) concepts. The arguments of Parmenides, Zeno's paradoxes about points and lines, subdivision, parts and wholes, and the arguments Plato develops in his dialogue *Parmenides* show what wonderful dream castles could be built from ideas no longer contaminated by the idiosyncracies of the particular. But the fact that simple ideas can be connected in simple ways gives the resulting propositions special authority only if everything can be shown to consist of simple things – which was precisely the point on which dis-agreements arose! 'We don't deal with Being, we deal with milk, pus, urine!' said some early physicians in a criticism I shall quote presently. The authority of the new enterprise, therefore, did not lie in the ideas and their connections themselves, but in the decisions of those who preferred neat constructions to analogies, who, like Parmenides, were not overly interested in crude empirical matters and who objectivised their lack of interest by saying that such things were not real (Parmenides B2,6): *the discovery of proof-procedures increased cultural variety, it did not replace it by a single true story*. This is confirmed by the whole history of Western thought.

---

37. This becomes clear from his identification of Thought and Being (Diels-Kranz, fragment B3). The identification was customary in archaic Greece but Parmenides was the first to use it as an argument against cultural opportunism.

A modern version of the belief that some traditions are not just better than the rest but are of an entirely different kind and that they alone give us knowledge is found in B.L. van der Waerden, *Science Awakening*, New York 1963, p. 89. Van der Waerden describes the different ways in which Babylonian and Egyptian mathematicians computed the area of a circle and he asks: 'How was Thales to discriminate between the exact, the correct recipes for computation and the approximate, the incorrect ones? Obviously by proving them, by fitting them into a logically connected system!'

Those who followed Parmenides were the first to relapse. They readmitted commonsense, though hesitatingly, and in small doses. The atomists, Empedocles, and Anaxagoras all accepted Parmenides' idea of Being but they also tried to retain change. To achieve their aim they introduced a (finite or infinite) number of things, each possessing some Parmenidean properties: the atoms of Leucippus and Democritus were indivisible and permanent, but infinite in number; the elements of Empedocles finite in number, permanent, divisible into regions, but not divisible into further substances (the four elements of Empedocles, the Hot, the Cold, the Dry and the Moist, were therefore different from any known substance); while Anaxagoras assumed the permanence of all substances. (Philosophical) theory was now a little closer to experience – but the remaining distance from commonsense and from the sciences of the time was still enormous.

Others had no compunction about rejecting the whole approach. Thus the author of the treatise *Ancient Medicine* not only used experience as a matter of course but also ridiculed those who, like Empedocles, had tried to replace it by more abstract considerations. 'I am at a loss to understand', he writes in chapter 15,

> how those who maintain the other view and abandon the old method in order to rest the *techne* on a postulate [i.e. who introduce theoretical principles] treat their patients on the lines of this postulate. For they have not discovered, I think, an absolute cold and hot, dry and moist [Empedoclean elements] that participates in no other form. But I think they have at their disposal the same foods and the same drinks we all use, and to the one they add the attribute of being hot, to another, cold, to another, dry, to another, moist, since it would be futile to order a patient to take something hot, as he would at once ask 'what hot thing?' So they must either talk nonsense, or have recourse to one of the known substances.

The difference between theoretical speculation and the empirical knowledge assembled by the practitioners of medicine cannot be described more clearly. In the quotation the philosopher is Empedocles with his four abstract substances. To explain the matter more directly, let us take Thales. Thales, according to tradition, had only one element, water. Hence, the only advice a Thalesian doctor could give a patient was either 'take water' or 'don't take water'. This is obviously 'nonsense' – see the above quotation. A doctor must specify, he must tell a patient 'what watery thing' he should consume or avoid. For example, he must say: 'take a little bread dipped in milk' or 'avoid wine at all costs, drink lukewarm cider, lots of it' – and so on. He must refer 'to the same food and the same drink we all use' and give his prescriptions as experience and the tradition of his craft have taught him. Being a Thalesian he might expand every prescription by adding 'and this is water, according to the most recent advances of natural philosophy' – empty words, to say the least.

The concept of health is empirical, or 'historical', to an even larger extent. It contains what happened to generations of patients and physicians side by side with their ideas of what a good life is supposed to be. It depends on the customs of those desiring health, it changes in time and it cannot be summarised in a definition. Empedocles did give a definition. Health, he said, is the balance of the elements (i.e. of his abstract substances) in the human body, illness their imbalance. This increased the number of ideas about health, it did not reduce them to one. Besides, practicing physicians rejected the definition out of hand. It 'no more pertains to medicine than to painting', wrote the author of *Ancient Medicine* (chapter 20).

The author of *Ancient Medicine* and other early opponents of the excesses of the theoreticians (Herodotus is an example) expressed their objections in writing – they were members of a tradition of written exchange that was soon to dominate Western civilization. Not all crafts participated in this tradition; we have no written reports from potters, metal workers, architects, miners, painters. We must reconstruct their knowledge from their work and from indirect references to it. Cyril Stanley Smith, a metallurgist from MIT, did this in a book as well as in an exhibition.[38] Like Herodotus (cf. his criticism of earlier geographical descriptions) and the author of *Ancient Medicine*, he distinguishes between philosophical theories (of matter) and a practical knowledge (of materials). He describes how the latter arose millenia before the former and was often impeded by them (for example in the neglect of alloys by the believers in Dalton's theory during the nineteenth century) and how it merged with them in the twentieth century, after physics had modified its views of reality. Norma Emerton describes the battle between

38. *A Search for Structure*, Cambridge, Mass. 1981. Photographic displays from the exhibition with analyses are published in *From Art to Science*, MIT Press 1980. Older works are V. Gordon Childe, *The Prehistory of European Societies*, Harmondsworth 1958 and C. Singer, E.J. Holmyard, and A.R. Hall, eds., *A History of Technology*, Vols. 1 and 2, Oxford 1954, 1956.

form theories (which were fairly close to the practice of the crafts) and atomism (which was not) and comments on the methods the atomists used to remain on top.[39] Altogether it turns out that technology, large parts of medicine, agriculture, and a practical knowledge of plants, animals, humans, societies, and even of the social dangers of knowledge (cf. the comments in footnote 9) owe much less to theoretical speculation than modern defenders of basic science claim, and were often hindered by it.[40]

Democritus' atomism did not add to knowledge; it was parasitic on what others had found in a non-theoretical way, as Democritus himself concedes (Diels-Kranz, fragment B 125).

The clearest objections to the Parmenidean approach came from the sophists and from Aristotle. Parmenides had thought that argument was a transtraditional means of finding truth. The sophists objected that a truth that is not part of a tradition is an impossibility; it cannot be found, if found it cannot be understood, and if understood it cannot be communicated. '*Being* is unknown', said Gorgias (Diels-Kranz, Fragment B 26), 'unless it *appears* in opinion' (my emphasis).[41] Commenting on Platonists who justified the virtues by reference to a supreme Good, Aristotle wrote (*EN* 1096b33ff, my emphasis):

> Even if there existed a Good that is one and can be predicated generally or that exists separately and in and for itself, it would be clear that such a Good can neither be produced nor acquired by human beings. *However it is just such a Good that we are looking for* . . . one cannot see what use a weaver or a carpenter will have for

39. Norma E. Emerton, *The Scientific Reinterpretation of Form*, Cornell University Press 1984. Cf. also the review by C.S. Smith in *Isis*, Vol. 76 (1985), pp. 584ff, esp. p. 584: 'One can see that today's quantum and solid state theories are part of a magnificent chain of arguments about the primacy of form or matter that started with Plato and Aristotle and moved from practical (empirical) to highly theoretical notions and back again.'

40. They were part of the very same 'empirical and unsystematic' traditions which von Hayek, in his discussion of liberty, contrasts with the speculations of (philosophical or scientific) system-builders: *The Constitution of Liberty*, p. 54.

41. Here the word 'appear' must not be interpreted in too narrow a manner. For example, it must not be interpreted as implying a naive sense-datism. 'To appear in opinion' simply means: to be part of some tradition. Arguments to the effect that nothing exists, that if anything existed it could not be found, that if it were found it could not be understood and if understood, not communicated were provided by Gorgias in his treatise *On the Non Existent* or *On Nature*. The arguments establish that a Being that exists independently of a tradition is a non-entity. Cf. also section 5. The sophists were the first (in the West) to be aware of the close connection between Being and Opinion.

his own profession from knowing the Good in itself or how somebody will become a better physician or a better general once 'he has had a look at the idea of the Good' [apparently an ironical quotation of a formula much used in the Platonic school]. It seems that the physician does not try to find health in itself, but the health of human beings or perhaps even the health of an individual person. For he heals the individual.

Aristotle also pointed out that 'natural things', i.e. the things that occur in our lives, 'are some or all of them subject to change' (*Physics*, 185a12f): a particular mode of existence, the waking state of a healthy human being, is made the measure of truth and reality.

This is a most interesting procedure. Aristotle does not produce an internal criticism of Parmenides' reasoning (he has such arguments, too, but they do not concern us here); nor does he compare it with abstract principles of his own. *He rejects the entire approach.* The task of thought, he seems to say, is to comprehend and perhaps to improve what we do when engaged in our ordinary everyday affairs; it is not to wander off into a no-man's land of abstract and empirically inaccessible concepts. We have seen that practitioners of the crafts who left records held similar opinions. I shall now give two examples to show that Greek commonsense concurred, not in argument, but by simply disregarding attempts at a theoretical reform.

My first example comes from *theology*. The Homeric gods were a mixed lot, but they all had human characteristics. They entered human lives, they were not merely postulated, they were seen, heard, felt, they were present everywhere. The day-to-day activities of the Greek tribes and even of an 'enlightened' city culture such as fifth-century Athens were organised around them.[42] It mattered little to the members of this rich and complex way of life that Xenophanes, using a drastically reduced notion of divinity, had proved that there was only one god, that he (it?) was devoid of human frailty but full of intelligence and power, and that the conventional gods were too approachable to be divine. Xenophanes' mockery of the Homeric gods impressed neither popular piety, nor such enlightened thinkers as Herodotus and Sophocles; even Aeschylus, who took over some

42. For details and literature cf. chapter 17 of my book *Against Method*, London 1975. The role of religion in fifth-century Athens is described in T.B.L. Webster, *Athenian Culture and Society*, University of California Press 1973, ch. 3.

Xenophanean formulae, still retained the traditional gods and most of their functions. The battle between theologians who conceive of God(s) in theoretical terms and dabble in proofs and the proponents of a personal or 'empirical' religion has lasted until today.

The second example is the failure of philosophers to make the use of *general concepts* a popular habit. Knowledge, in tradition and in Greek commonsense, was a collection of opinions, each of them obtained by procedures appropriate to the domain from which the opinions arose. The best way of presenting such knowledge is the list – and the oldest scientific works were indeed lists of facts, parts, coincidences, problems in various and occasionally already specialised domains. The answers the Platonic Socrates gets to his inquiries show that lists were also part of commonsense. His objection 'I asked for one and I get many' assumes that one word means one thing – the point at issue. His interlocutors grant unity to numbers (*Theaetetus*) or bees (*Meno*), but balk at extending theoretical uniformity to social affairs such as knowledge and the virtues: Plato was well aware of the difficulty of extending simple concepts to complex matters. This issue, too, has remained alive until today – as the rift between the sciences and the humanities.

The case of *mathematics* is especially interesting. It was here that abstract thought first produced results and it was from here that the paradigm of true, pure and objective knowledge spread to other areas. But the many approaches mathematics now contains show no tendency to coalesce into a single theory. We have non-Euclidian geometries and various versions of arithmetic; finitists regard mathematics as a human practice which, depending on the aim, can be built up in different ways; 'Cantorians' interpret it as a science descriptive of abstract entities and therefore in need of unity; the application of a particular mathematical system to 'nature' recreates the plurality (of approximations) Thales allegedly removed (cf. footnote 37); new mathematical disciplines arise all over the place. Today mathematics is less restrained and more pluralistic than any other intellectual discipline.

These historical results can be summarised in the following statement:

R8: the idea of an objective truth or an objective reality that is independent of human wishes but can be discovered by human effort is part of a special tradition which, judged by its own members, contains successes as well as failures, was always accompanied by, and often mixed with, more practical (empirical, 'subjective') traditions, and must be combined with such traditions to give practical results.

R8 is an empirical (historical) thesis. An empiricist infers

R9: the idea of a situation-independent objective truth has limited validity. Like the laws, beliefs, customs of R4 it rules in some domains (traditions), but not in others.

This strengthens R7 and the considerations of the previous section. Note again that R8 and R9 are not 'universal truths'; they are statements which I, as one member of the tribe of Western intellectuals, present to the rest of the tribe (together with appropriate arguments) to make them doubt the objectivity and, in some forms, also the feasibility of the idea of objective truth.

7 Epistemic Relativism

R8 and R9 deny that the new forms of knowledge that arose in Greece, and later on led to the sciences, can overrule (and not merely overrun) traditions and establish a tradition-independent point of view. The reasons I gave for the denial were partly historical, partly anthropological: opinions not tied to traditions are outside human existence, they are not even opinions while their content depends on, or is 'relative to', the constituting principles of the traditions to which they belong. Opinions may be 'objective' in the sense that they do not contain any reference to these principles. They then *sound* as if they had arisen from the very essence of the world while merely reflecting the peculiarities of a particular approach: the values of a tradition recommending absolute values may be absolute, but the tradition itself is not; physics may be 'objective' but the objectivity of physics is not. More recently, objectivist traditions have produced points of view that do not even sound objective. The theory of relativity asserts the relational character of situations and events which a century ago were regarded as existing independently of measurement, while the quantum theory in addition lacks the invariants which still enable us to objectivize relativity. Also the objectivist tradition has long ago split into competing schools or, in the case of the sciences, into approaches based on different assumptions and employing different methods. Unpopular and even 'untenable' ideas have entered it and have become the law of the land, successful principles have been overtaken and relegated to the garbage heap of history. Developments such as these (and the additional remarks in footnotes 16 to 19 above) suggest the following hypothesis:

R10: for every statement (theory, point of view) that is believed to be true with good reasons *there may exist* arguments showing that either its opposite, or a weaker alternative is true.

We can go still further. I mentioned in the preceding section that ancient arguments against Parmenides' monism contained two steps: the decision to keep close to experience; and theoretical considerations building on this decision. Herodotus already knew that there exist different ways of arranging experience, each offering its own account of the world and its own ways of dealing with it. He also knew that people not only live in these different worlds, they live successfully, both in the material and in the spiritual sense. Modern anthropologists agree. 'Let the reader consider any argument that would utterly demolish all Zande claims for the power of [their] oracle[s]', writes E.E. Evans Pritchard, reporting a case I mentioned in the introduction.[43] 'If it were translated into Zande modes of thought it would serve to support their entire structure of belief. For their mystical notions are eminently coherent, being interrelated by a network of logical ties and are so ordered that they never too crudely contradict sensory experience but, instead, experience seems to justify them.' Result: Zande practices are 'rational' because supportable by argument. They also work. 'I may remark', writes Evans-Pritchard on this point, 'that I found this [i.e. consulting oracles for day-to-day decisions] as satisfactory a way of running my home and affairs as any other I know of.'

Adding the points made in the literature cited in footnote 9, we arrive at the hypothesis that there exist many different ways of

---

43. *Witchcraft, Oracles and Magic Among the Azande*, Oxford 1973, pp. 319f. The second remark is from p. 270. Oracles have many advantages over 'rational discussions'. They don't exhaust those who use them and they make it clear that the consultants have no say in important matters. An extended rational discussion, on the other hand, may be so chaotic and tiring that it turns its participants into mere randomizers. They act then like oracles, but without strength, and in the firm belief that they are still the masters of their fate.

living and of building up knowledge. Each of these ways may give rise to abstract thought which in turn may split into competing abstract theories. Scientific theories, to give an example from our own civilization, branch out in different directions, use different (and occasionally 'incommensurable') concepts and evaluate events in different ways. What counts as evidence, or as an important result, or as 'sound scientific procedure', depends on attitudes and judgements that change with time, profession, and occasionally even from one research group to the next. Thus Ehrenhaft and Millikan working on the same problem (the charge of the electron) used their data in different ways and regarded different things as facts. The difference was eventually removed but it was the core of an important and exciting episode in the history of science. Einstein and the defenders of hidden variables in the quantum theory use different criteria of theory evaluation. They are metaphysical criteria in the sense that they support or criticize a theory although it is empirically satisfactory and well formulated mathematically.[44] The same is true of criteria that stretch an empirical subject beyond the reach of its evidence, asserting, for example, that all biology is molecular biology and that botany has no longer any independent claim to truth. T.H. Morgan, preferring direct experimental support to data involving inferences, rejected the study of chromosomes in favour of more overt manifestations of inheritance. In 1946 Barbara McClintock had already noticed the process which today is called transposition. 'However she worked alone, she did not work with microorganisms, she worked in the classical manner and stayed away from molecules.' Not a single member of the rapidly expanding group of molecular biologists 'listened to what she said'. Divergences proliferate in psychology: behaviourists and neurophysiologists despise introspection which is an important source of knowledge for gestalt psychology, clinical psychologists rely on their experience, sometimes called 'intuition', i.e. on the reaction of their own well prepared organism, while more 'objective' schools use rigorously formulated tests instead. In medicine a similar antagonism between clinicians and body theoreticians goes back to antiquity, as we have seen. Differences increase when we move on to history and sociology: a social history of the French Revolution shares only names with

44. Similarly Copernicus criticized the existing planetary theories while admitting that they were all 'consistent with the data'. *Commentariolus*, quoted from E. Rosen, ed., *Three Copernican Treatises*, New York 1959, p. 57.

a description of persons and concrete individual events.[45] Nature herself can be approached in many ways (the idea that there exists no separation between her and the lives of humans being one of them, the idea of her non-material character being another) and responds accordingly. Taking all this into account I suggest that we strengthen R10 and assert

> R11: For every statement, theory, point of view believed (to be true) with good reasons *there exist* arguments showing a conflicting alternative to be at least as good, or even better.

R11 was used by the ancient sceptics to achieve mental and social peace: if opposing views can be shown to be equally strong, they

45. The case of Ehrenhaft and Millikan is discussed in G. Holton, *Historical Studies in the Physical Sciences*, Vol. ix, R. McCormmich, L. Pyenson and R.S. Turner (eds.), Johns Hopkins University Press 1978, pp. 161–214. For T.H. Morgan's objections to the Sutton-Boveri chromosomal theory of inheritance cf. E. Mayr, *The Growth of Biological Thought*, Cambridge 1982, 748f. Mayr's book contains many examples of the way in which different research traditions using different evidence may come to different conclusions about what they regard, vaguely, as 'the same thing'. Mayr therefore objects to interpreting the history of science as a sequence of uniform paradigms (*op. cit.*, p. 113). The quotation about McClintock is from Peter Fischer, *Licht und Leben*, Konstanz 1985, p. 141. Cf. also E. Fox-Keller, *A Feeling for the Organism*, San Francisco 1983. The clinical-statistical controversy is surveyed in Paul E. Meehl, *Clinical vs. Statistical Predictions*, Minneapolis 1954. For a wider discussion of 'objective' measures of human worth cf. R.C. Lewontin, S. Rose and L.J. Kamin, *Not in Our Genes*, New York 1984 as well as St. Gould, *The Mismeasure of Man*, New York 1981. Stanley Joe Reiser, *Medicine and the Reign of Technology*, Cambridge 1978, contains a discussion of the changing antagonism between healers who directly inspect the human body and body theoreticians who favour 'objective' tests as described in Norma E. Emerton, *The Scientific Reinterpretation of Form*, Cornell University Press 1984, Cf. also chapter 5 of C.S. Smith, *A Search for Structure*, MIT Press 1981.

Ilya Ehrenburg (*People and Life*, Memoirs of 1891–1917, London 1961, p. 8) writes as follows about the 'French Revolution' (retranslated from the German): 'The images which authors hand over to later generations are formalised and occasionally completely contrary to truth . . . There is sometimes talk about the "Storming of the Bastille" though in reality the Bastille was not stormed by anyone – 11 July 1789 was merely one episode in the French Revolution; the people of Paris entered the prison without difficulty and found there only a few prisoners. But just this capture of the Bastille became the national holiday of the revolution.' (Compare with this the emphasis on special dates, events, 'discoveries' in the history of science and in the the preparation of Nobel Awards.) For the astounding difference between standardised (or streamlined) versions and 'actual events' cf. Georges Pernoud and Sabine Fleissier, *The French Revolution*, New York 1960. Eisenstein when preparing his *Potemkin* knew very well that history had to be improved in order to become exciting and meaningful. Lakatos realised the same for the sciences.

said, then there is no need to worry, or to start a war about them (Sextus Empiricus, *Hypot.*, 1, 25f). Statements, theories, arguments, good reasons enter the scene because of the historical situation in which the sceptics made their point: they opposed philosophers who had tried to show that argument would lead to unique conclusions; but argument, the sceptics maintained, has no such power. Including non-argumentative ways of establishing human contact and, possibly, a common purpose further strengthens their position. For now we are dealing not merely with intellectual matters, but with feelings, faith, empathy and many other agencies not yet catalogized and named by rationalists. A removal of R11 would require detailed empirical/conceptual/historical analyses none of which are found in the customary objections to scepticism and relativism.

8 Some Critical Remarks Examined

Relativism is a popular doctrine. Repelled by the presumption of those who think they know the truth and having witnessed the disasters created by attempts to enforce a uniform way of life many people now believe that what is true for one person, or one group, or one culture need not be true for another. This practical relativism is supported by the pluralism inherent in modern societies and especially by the discoveries of historians and anthropologists: ancient ideas and the 'primitive' cosmologies of our own times may differ from what we are accustomed to, but they have the capacity of creating material and spiritual well being. They are not perfect – no world view is – but their drawbacks, judged by our own ways of life, are often balanced by advantages we lack. Evolution provides still another argument: each section, division, phylum, species developed its own way of being in a world which is largely of its own making, with appropriate sense organs, interpretative mechanisms, ecological niches.[46] The world of a spider has little in common with the world of a dog and it could be quite silly of a canine philosopher to insist on the objective validity of his ideas. The ancient sceptics and their modern followers (Montaigne, for example) made excellent use of this variety.

On the other hand there are many people who regard such a situation as unsatisfactory and who try to find the single truth which, in their opinion, must be hidden beneath what otherwise would be a chaotic mass of information. Strangely enough there are relativists who share these aspirations. They do not merely want to air their own opinions about the efforts and products of traditions untouched by Western rationalism, they want to make general and – god help us! – 'objective' statements about the nature of knowledge and truth.

But if objectivism while perhaps acceptable as a particular point of view cannot claim objective superiority over other ideas, then the objective way of posing problems and presenting results is not the right way for the relativist to adopt. A relativist who deserves his name will then have to refrain from making assertions about the nature of reality, truth and knowledge and will have to keep to specifics instead. He may and often will generalise his findings but without assuming that he now has principles which by their very nature are useful, acceptable and, most importantly, binding for all. Debating with objectivists, he may of course use objectivist methods and assumptions; however his purpose will not be to establish universally acceptable truths (about particulars or generalities) but to embarrass the opponent – he is simply trying to defeat the objectivist with his own weapons. Relativistic arguments are always *ad hominem*; their beauty lies in the fact that the *homines* addressed, being constrained by their code of intellectual honesty, must consider them and, if they are good (in their sense), accept them as 'objectively valid'. All my arguments in the preceding sections should be read in this manner.[47]

46. In classical Darwinism the organisms adapt to a world that is given independently of their actions. This 'simple view that the external environment changes by some dynamic of its own and is tracked by the organism takes no account of the effect organisms have on the environment. . . . the organism and the environment are not actually separately determined. The environment is not a structure imposed on living beings from the outside but is in fact a creation of those beings.' R. Levins and R. Lewontin, *The Dialectical Biologist*, Cambridge, Mass. 1985, pp. 69, 99. For details consult chapters 2 and 3 of the book.

47. Hume, *A Treatise of Human Nature*, Book i, Of the Understanding, D.G.C. Macnabb, ed., New York 1962, pp. 236f describes the situation as follows: 'Reason first appears in possession of the throne, prescribing laws, and imposing maxims, with an obsolute sway and authority. Her enemy, therefore, is obliged to take shelter under her protection and, by making use of rational arguments to prove the fallaciousness and imbecility of reason, produces, in a manner, a patent under her hand and seal. This patent has at first an authority, proportional to the present and immediate authority of reason, from which it is derived. But as it is supposed to be contradictory to reason, it gradually diminishes the force of the governing power and its own at the same time; till at last they both vanish away into nothing, by a regular and just diminution.'

For example, R7 to R11 are not meant to reveal 'objective features' of the world; they are introduced to undermine the objectivist's confidence or to capture outsiders by a vivid historical image.[48] If the objectivist agrees with my arguments, then R1 and R7 become difficulties for his point of view and this quite independently of whether I myself believe in them or not. I shall now apply this procedure to some popular objections against relativism.

The first objection, heard rather frequently, is not so much an objection as a curse. 'Relativism', says Karl Popper (*Auf der Suche nach einer besseren Welt*, Munich 1984, p. 217), 'is the position that anything can be asserted, or almost anything and therefore nothing . . . Truth, therefore, is meaningless.' Relativism 'comes from a lax tolerance and leads to the rule of power'.

The quotations in section 3 (Herodotus and Protagoras) show that the first part of the curse and its end ('lax tolerance') are both incorrect. Herodotus (whom Popper quotes on page 134 of his book, carefully omitting the lines that undermine his travesty of relativism) was a relativist; so was Protagoras. But the former emphasized and defended the power of customs while the latter recommended death for repeated violators of the law. 'As can be seen,' I wrote in that section, 'Protagoras believed that there had to be laws and that they had to be enforced. He also believed that laws and institutions had to be defined "relative to" the needs and the circumstances of these societies. Neither he nor Herodotus inferred . . . that institutions and laws that were valid in some societies and not valid in others were therefore arbitrary and could be changed at will.' Popper's further accusation that 'truth . . . is meaningless' for relativists clashes with the careful way in which Protagoras discusses the use of this term.

In an addendum to Vol. 2 of his *The Open Society and its Enemies* (New York 1966, pp. 369ff), Popper explains his attitude in greater detail. He starts with a definition: 'By relativism – or, if you like, scepticism – I mean . . .' Note the nonchalant 'if

48. This is my intention. Of course, many readers will find fault with my arguments. But they cannot criticize me for assuming objective principles as they do. I may use the principles badly; I may use the wrong principles; I may draw the wrong conclusions from them; but I intend to use them as rhetorical devices, not as objective foundations of knowledge and argument. The rationalist, on the other hand, being addressed in the (for him) appropriate way, will have to read my reasons 'objectively' and will thus end in confusion.

you like': there is no difference between scepticism and relativism in Popper's mind. But there is a great difference in history. The sceptics offered a diagnosis of their times, an aim for philosophers to pursue, and an argument. Their aim was peace, their diagnosis that quarrels about abstract dogmas can lead to dissension and war, and their argument that any well supported statement could always be balanced by an equally well supported opposite. The aim is admirable, the diagnosis still correct, the argument detailed and, as I tried to show in the preceding section, rather strong. None of these features surfaces in Popper's definition.

According to Popper, relativism ('or, if you like, scepticism') is 'the theory that the choice between competing theories is arbitrary; since either there is no such thing as objective truth; or, if there is, no such thing as a theory which is true or at any rate (though perhaps not true) nearer to the truth than another theory; or, if there are two or more theories, no ways or means of deciding whether one of them is better than another.'

Note again that both the first statement (arbitrariness of choice) and the last one (no means of deciding between alternative views) conflict with what Plato tells us about Protagoras and that the ancient sceptics offered arguments for the first statement. Critics of the idea that scientific debates are settled in an objective manner do not deny that there exist 'means of deciding' between different theories. On the contrary, they point out that there are many such means; that they suggest different choices; that the resulting conflict is frequently resolved by powerplay supported by popular preferences, not by argument; and that argument at any rate is accepted only if it is not just valid, but also plausible, i.e. in agreement with non-argumentative assumptions and preferences.

Popper calls relativism a 'theory'. This covers some versions, but omits others (my own included), as we have seen. He identifies the problem of (the objectivity of) *knowledge* with the problem of the truth and/or objectivity of *theories*. This may work for parts of physics (though there is 'tacit knowledge' even here) but it is much too narrow an approach for history, psychology and for the wide domain of commonsense.

'If two parties disagree,' says Popper (p. 387), 'this may mean that one is wrong, or the other, or both. It does not mean, as the relativist will have it, that both may be equally right.'

This comment reveals in a nutshell the weakness of all intellectual attacks on relativism. 'If two parties disagree' – this means the opponents have established contact and understand each other. Now assume that the opponents come from different cultures. Whose means of communication will they use and how will understanding be reached? Colonial officials took it for granted that the natives would either learn the Master Language or could be informed by interpreters, again using the Master Language as a basis. The Master Language, applied in situations defined by the Masters, was the official medium of formulating, presenting and solving problems. Can we take it for granted that using indigenous means of establishing contact, an indigenous language and indigenous ways of solving problems would have led to the same solutions? To the same problems? Older studies and the recent experiences of professional 'developers' caution us against such an assumption. But then the disagreements encountered and the divisions into right and wrong they engender depend on the form of the interaction, hence on the culture; they are 'relative to' the culture in which the exchange takes place. Popper, like some minor lights of the Enlightenment before him, seems to assume that there exists, basically, a single medium of discourse that the medium is 'rational' in his sense (for example, it obeys simple logical laws), that it consists mainly of talk (gestures, facial expression play no role), and that everybody has access to it.

> Und unterm braunen Sud fühlt auch der Hottentot
> Die allgemeine Pflicht und der Natur Gebot

wrote Albrecht von Haller,[49] turning everybody into a potential Kantian. In the same way Popper perceives a tiny and somewhat confused Popperian behind every human face and he sternly criticizes people for giving in to their confusion. Moreover, he misses the point of relativism even in this already quite narrow domain: Protagoras would not have called conflicting positions 'equally right'.[50]

Finally, why should it not be possible to say conflicting things about 'the same situation' and yet be right? A picture that can be seen in two different ways (Wittgenstein's duck-rabbit is an example) can be described in two different ways – and both parties will be right. It is a matter of research and not of philosophical fiat to decide whether the world we inhabit resembles a duck-rabbit picture.

Another critic of relativism is Hilary Putnam. In his book *Reason, Truth and History* (Cambridge 1981, p. 114), he writes:

'I want to claim that *both* of the two most influential philosophies of science of the twentieth century, certainly the two that have interested scientists and non-philosophers generally, the only two the educated general reader is likely to have heard of, are self refuting.' The philosophies he means are positivism as represented by Karl Popper and the historical approach as represented, among others, by Kuhn, Foucault and myself. To establish his claim he discusses incommensurability and relativism. Incommensurability will be dealt with in chapter 10. Here I want to examine Putnam's treatment of relativism.

Putnam starts with a version of relativism according to which 'no point of view is more justified or right than any other' (p. 119); he criticizes it by asking how one can *hold* that there is no reason for holding one point of view rather than another. The answer is simple: I can hold opinions without either having or giving a reason. Besides, the version is not my version.[51]

Next Putnam discusses what one might call 'relational' relativism: 'true', or 'reasonable', or 'acceptable' are to be replaced by 'true for', 'reasonable according to such-and-such criteria', 'acceptable for a member of culture A' – and so on. 'A total relativist' (in this sense), says Putnam (p. 121), 'would have to say that whether or not X is true *relative to* P is *itself* relative. At this point our grasp on what the position even means begins to

---

49. *Über den Ursprung des Übels* (1750 ed.), Vol. 2, p. 184: 'And under his brown skin even the Hottentott has a feeling for universal duty and the commandments of nature'. Cf. A. Lovejoy, *Essays in the History of Ideas*, Baltimore 1948, pp. 78ff, esp. 86f.

50. Like his other achievements, Popper's error is not new. It is found in Plato and Aristotle. Plato starts one of his arguments against Protagoras with the following version of Protagoras' doctrine (*Theaet.*, 170a3f): 'What everyone thinks, he says, is *for him* who thinks it'. He points out that few people are ready to accept this doctrine. Most people rely on experts. Truth, *for them*, is what is provided by experts. Thus Protagoras, having made opinion the measure of truth and existence, must admit that his doctrine is *false*. End of argument. The argument involves a transition from 'being for' or 'true for' to 'being' and 'true' and is therefore a non sequitur.

Aristotle, discussing the principle of non-contradiction, enumerates philosophers who violate it. 'The saying of Protagoras', he writes (*Met.*, 1062b13f), 'is like the view we have mentioned; he said that man is the measure of all things, meaning simply that this which seems to each man assuredly is. If this is so, it follows that the same thing both is and is not, and is bad and good, and that the contents of all other opposite statements are true.' But, as Plato clearly says, Protagoras identifies what is thought by a person with what is *for that person*, and not absolutely.

51. I explicitly reject it in *Science in a Free Society*, London 1978, p. 83.

wobble . . .' It certainly does – but only if the 'position' is read as an objective account of knowledge. A rhetorical account that addresses objectivists with the intention of confusing them is already talking to the right party and can therefore omit the 'for'.

Putnam also asserts (p. 122) that a culture that does not distinguish being from seeming cannot separate asserting (thinking) from making noises and is therefore no longer a culture. This is a good example of the abstract way in which philosophers deal with problems of living. As I pointed out at the beginning of section 6, there exist many forms of life, Homeric commonsense among them, that lack gross dichotomies such as the dichotomy being-appearing. Being aware of the complexity of the world and of human action they use a variety of subtle distinctions instead. Putnam, applying his crude conceptual grid, is forced to reject much of their talk as mere noise. This is a criticism of the grid, not of the talk rejected. Besides, relativists would not be defeated by the distinction for they can point out (cf. again section 6) that different cultures and even different schools within one culture draw the line in different places.

Return to Life

In conclusion let me repeat that relativism as presented here is not about concepts (though most modern versions of it are conceptual versions) but about human relations. It deals with problems that arise when different cultures, or individuals with different habits and tastes, collide. Intellectuals are accustomed to deal with cultural collisions in terms of debates and they tend to refine these imaginary debates until they become as abstract and inaccessible as their own discourse. Proceeding in this manner many of them have moved away from life into a realm of technical knowledge. They are no longer concerned with this or that culture or this or that person; they are concerned with ideas such as the idea of reality, or the idea of truth, or the idea of objectivity. And they do not ask how the ideas are related to human existence, but how they are related to each other. For example, they ask if truth is an objective notion, if scientific practice is rational, or how reality depends on perception – where 'truth', 'scientific practice' and 'perception' are defined in ways that prevent a ready identification with what goes on in the lives of scientists and other ordinary human beings (cf. section 4 for details).

Entire professions are devoted to clarifying questions of this kind. By now the resulting word games have become a world-wide malaise. They are played by Western intellectuals; they have also caught the attention of non-Western observers fascinated by the brazen splendour of the products of Western Civilization. Confounding the intellectual power of ideas with the political and military power of the societies containing them, men and women from the so-called 'Third World' have started immersing themselves in the mudbath of Western philosophy. But the entire development, far from starting a Renaissance of thought, has only discredited it; it has led to what some philosophers, being unable to look beyond their own playpens, have called a 'Crisis of World Culture'. I, on the other hand, believe that the crisis lies not in intellectual and academic life but in large scale phenomena intentionally or inadvertently supported by the products of this life. To reveal the support we must identify the hidden assumptions and the gross errors behind the alleged objectivity of Western intellectual products. But having revealed them it is equally important to return to life and to deal with its problems in a more direct way, for example by studying the reactions of individuals and societies confronted with unusual situations.

As I pointed out in the Introduction, clashes between cultures lead to a variety of reactions. One of the reactions is *dogmatism*: our way is the right way, other ways are false, wicked, godless. Some dogmatists are tolerant – they pity the godless, try to inform them, but otherwise leave them alone. The tolerance of some sixteenth and seventeenth century Christians was of this kind. Others fear that the proponents of falsehood might corrupt Truth, and suggest killing them. This was the point of view of Deuteronomy. Modern dogmatists, living in democracies where pluralistic and libertarian rhetoric prevails, seek power in a more underhanded way. Distinguishing between 'mere beliefs' and 'objective information', the defenders of scientific rationalism tolerate the former but use laws, money, education, PR to put the latter in a privileged position. They have succeeded to a surprising extent. The separation of Church and State, laws prohibiting all but officially recognized medical procedures, strict educational policies, the combination of science with nationally important projects such as national defence — all tend to strengthen what powerful groups regard as objective truth and to weaken opinion.

In the preceding sections I tried to show that dogmatism led to disastrous consequences when used as a principle of cultural exchange and/or cultural growth. Even Western observers now admit that something went wrong when Western technologies and Western ways of living were transferred to regions as yet untouched by Western history. Life in these regions was not perfect; it had large lacunae (no effective treatment for many diseases) and contained ingredients inimical to well-being. In this respect it was rather similar to what we have now in the West. But the wholesale removal of traditional customs and their wholesale replacement by 'rational' procedures was not the right solution. The 'right' solution, many people now suggest, is to take both local and Western knowledge into account and to utilize them in accordance with the customs of the affected communities. It is true that these customs are not always beneficial even when judged by those who practice them; but they are part of the lives of the people and therefore natural reference points. Disregarding them means treating people as slaves in need of instruction by superior masters.

The comments just made apply to forms of life that are explicitly dogmatic; however they also apply to philosophies that pride themselves on their modesty, their tolerance and their critical posture. At first sight such philosophies seem ideal instruments for cultural exchange. They concede that the teachers, the representatives of science and rationalism, may be mistaken and that the pupils, the representatives of an indigenous culture about to be introduced into Western ways, may have better things to offer. This seems to be a very tolerant and humane attitude indeed. It is tolerant – by 'critical' standards. For it assumes that the exchange will take the form of a debate, that the debate will be conducted according to certain rules and that its outcome decides the matter. It reduces human contact to verbal exchange and verbal exchange to debate, and it further reduces the debate to a search for the logical faults of clearly formulated issues. From the very beginning 'critical' philosophers define human relations in their own intellectualised way. Congratulating themselves on their tolerance they are either ignorant, or dishonest, or (my own conjecture) both.

Relativism moves away from this ignorance and dishonesty. It says that what is right for one culture need not be right for another (what is right for me need not be right for you). More abstract formulations which arose together with Western rationalism assert that customs, ideas, laws are 'relative to' the culture that has them. Relativism in this sense does not mean arbitrariness (this matter was discussed in section 3 and again in section 8) and it is not 'valid for' relativists only. Revealing major lacunae in the objectivist framework, it dissolves objectivism from the inside, according to the objectivists' own criteria.

Opportunism is closely connected with relativism; it admits that an alien culture may have things worth assimilating, takes what it can use and leaves the rest untouched. Opportunism played a large role in the spreading of Western science.

An episode in the history of Japan will illustrate the process. In 1854 Commander Perry, using force, opened the ports of Hakodate and Shimoda to American ships for supply and trade. This event demonstrated the military inferiority of Japan. The members of the Japanese Enlightenment of the early eighteen seventies, Fukuzawa among them, now reasoned as follows: Japan can keep its independence only if it becomes stronger. It can become stronger only with the help of science. It will use science effectively only if it does not just practice science, but also believes in the underlying ideology. To many traditional Japanese this ideology was barbaric (I would agree). But, so the followers of Fukuzawa argued, it was necessary to adopt barbaric ways, to regard them as advanced, to introduce the whole of Western civilization in order to survive. Note the strange but coherent reasoning: science is accepted as a true description of the world not because it *is* a true description, but because teaching it as such will produce better guns. The 'progress of science' would collapse without such events.[52]

Argument plays an important role in all forms of cultural exchange. It was not invented by Western rationalists. It occurs in all periods of history and in all societies. It is an essential part of the opportunistic approach: an opportunist must ask himself how foreign things are going to improve his life and what other changes they will cause. Occasionally 'primitives' used argument to turn the tables on anthropologists who tried to convert them to rationalism (see the example in the text to footnote 43). Argument, like ritual, or art, or language, is universal; but again like ritual, or language, or art, it has many forms. A gesture or a grunt may convince some participants while others need long and colourful arias. Luther wanted miracles from those proposing new interpretations of the holy texts; government institutions and the general public still want miracles from their own religious leaders, the scientists. Most arguments take the beliefs or the attitude of the participants into account. Those who use them want to persuade particular people and they change their approach from one case to the next. What the early Western rationalists did invent was not argument, but a special and standardised form of argumentation which not only disregarded but explicitly rejected personal elements. In return, the inventors claimed, they could offer procedures and results that were valid independently of human wishes and concerns.

In section 6 I explained why this claim was mistaken. The human element was not eliminated, it was only concealed. A colonial official spoke in the name of his king, a missionary in the name of God or of the Pope. Both could and did identify the authority that gave strength to their demands. Rationalists, too, have their authorities; but by speaking in an objectivist manner, carefully omitting any reference to the people they are trying to emulate and the decisions that made them adopt their procedures, they create the impression that Nature herself or Reason herself supports their ideas. A closer look at their procedure shows that this is not the case. Take success. Today the success of a procedure often counts as a sign of its objective validity. But the evaluation of successes and failures depends on the culture in which these events are taking place. Thus the so-called 'green revolution' was a success from the standpoint of Western marketing practices but a dismal failure for cultures interested in self-sufficiency. Moreover, there exists no 'objective' scientific study of the comparative effectiveness of Western and indigenous procedures in many fields. Even medicine can only offer isolated reports of successes, and equally isolated reports of the failures of non-Western medical practices: but the overall picture is far from clear.

A more sophisticated argument asserts that while success may be culture-dependent, the validity of the laws that are used to bring it about is not. People may differ in their attitudes towards electrification, but Maxwell's equations and their consequences are valid independently of these differences. This argument assumes that theories are not changed in application. But many so-called 'approximation procedures' remove what is asserted by the theory used and replace it by different assertions, thus admitting that different domains require different procedures and that the unity suggested by a comprehensive theory may be purely formal.

52. Details in Carmen Blacker, *The Japanese Enlightenment*, Cambridge 1969. For the political background cf. chapters 3 and 4 of Richard Storry, *A History of Modern Japan*, Harmondsworth 1982.

An even more important reply is that laws of nature certainly are not *found* independently of a particular culture. It needs a very special mental attitude inserted into a particular social structure combined with sometimes quite idiosyncratic historical sequences to divine, formulate, check and establish laws such as the second law of thermodynamics. This is now admitted by sociologists, historians of science and even by some philosophers. The Greeks had the mathematics and the intelligence needed to start the kind of science that developed in the sixteenth and seventeenth centuries – yet they failed to do so. 'Chinese civilization had been much more effective than the European in finding out about Nature and using natural knowledge for the benefit of mankind for fourteen centuries or so before the scientific revolution', and yet this revolution occurred in 'backward' Europe.[53] The discovery and development of a particular form of knowledge is a highly specific and unrepeatable process. Now where is the argument to convince us that what was found in this idiosyncratic and culture-dependent way (and is therefore formulated in culture-dependent terms) exists independently of the way that reached it? What guarantees that we can separate the way from the result without losing the result? If we replace some concepts by others, even only slightly different ones, we become unable to state these results, or even to comprehend them; we obtain different results and confirming evidence for them, as is seen when we move to earlier stages of the history of science. Yet the results are supposed to remain, 'in the world', long after we have forgotten how to reach them.

Moreover, modern objectivists are not the only people to project their fancies into the world. For the ancient Greeks, the Greek gods existed and acted independently of the wishes of humans. They simply 'were there'.[54] This is now regarded as a mistake. In the view of modern rationalists the Greek gods are inseparable parts of Greek culture, they were imagined, they did not really exist. Why the disclaimer? Because the Homeric gods cannot exist in a scientific world. Why is this clash used to eliminate the gods and not the scientific world? Both are objective in intention and both arose in a culture-dependent way. The only answer I have heard to this question is that scientific objects behave more lawfully than gods and can be examined and checked in greater detail. The answer assumes what is to be

53. J. Needham, *Science in Traditional China*, Harvard University Press/The Chinese University Press Hong Kong, 1981, pp. 3 and 22ff.

shown, namely that scientific laws are real while gods are not. It also makes accessibility and lawfulness a criterion of reality. This would make shy birds and anarchists very unreal indeed. There is no other way out: we either call gods and quarks equally real, but tied to different circumstances, or we altogether cease talking about the 'reality' of things and use more complex ordering schemes instead (cf. the beginning of section 6 above).

Neither option need affect the role of science in our culture. Nor am I asserting that we can do without the sciences. We cannot. Having participated in, or permitted, the construction of an environment in which scientific laws come to the fore, both materially, in technological products, and spiritually, in the ideas that are allowed to guide major decisions, we, scientists as well as the common citizens of Western civilization, are subjected to their rule. But social conditions change and science changes with them. Nineteenth century science denied the advantages of cultural plurality; twentieth century science, chastened by a series of rather upsetting revolutions and urged on by sociologists and anthropologists, recognizes them. The same scientists, philosophers, politicians who support science change science by this very support and they change the world with it. This world is not a static entity populated by thinking ants who, crawling all over its crevices, gradually discover its features without affecting them in any way. It is a dynamic and multifaceted entity which affects and reflects the activity of its explorers. It was once a world full of gods; it then became a drab material world and it will, hopefully, change further into a more peaceful world where matter and life, thought and feelings, innovation and tradition collaborate for the benefit of all.

54. For details and further literature, cf. chapter 17 of my book *Against Method*, London 1975. Cf. also chapters 4 and 5 of my *Stereotypes of Reality*, forthcoming.

2

Reason, Xenophanes and the Homeric Gods

Rationalism and science are conquering ever increasing sections of the globe. Education pounds them into the brains of the children of 'civilised' nations, development takes care that 'primitives' and 'underdeveloped' societies can profit from them, weapons research, which is an international enterprise and independent of political affiliations, introduces them to the very centres of power, even the smallest project has to be adapted to scientific standards to be acceptable. The trend has some advantages – but it has also serious drawbacks. 'Development', for example, often created the scarcity it is now trying to remove and destroyed institutions and cultures that sustained the lives of many people. Some critics have these drawbacks in mind when arguing against a further extension of the powers of science. They consider problems of *life*. They want to remove hunger, disease and fear, but they are aware of the dangers of science-based technologies; they work for peace and the independence of cultures different from our own and they deny that a scientific rationalism can achieve these aims.

There exist also other and more esoteric critics of the present trend. They hardly ever descend to such proletarian subjects as sanitation or the possibility of a nuclear war. They are not interested in the day-to-day existence of living things, of women, men, children, dogs, trees, birds. What they are concerned about is the power of special groups. This power, they point out, has suffered as a result of the expansion of the sciences. For example, the humanities now count for much less than the sciences and a thing called 'myth' has lost much of its influence. The criticism is followed by positive suggestions: give more money to the arts and the humanities and revive the mythical qualities of human life!

The suggestions assume a sharp distinction between pure thought with its artificial categories and myth, or the poetic imagination, which grasps human life as a whole and gives it meaning. In this assumption, the critics overlook that the distinction is itself a rational distinction. They criticize rationality on the basis of categories which were introduced by reason in the first place. Homer does not separate reason and myth, (abstract) theory and (empirical) commonsense, philosophy and poetry. Are the 'myths', is the 'poetic imagination' that modern esoteric thinkers have in mind perhaps mystifications, light years removed from the past they want to revive and the lives they want to enrich? And how does one react to the phenomena of this world and the opinions and institutions it contains without making use of the distinctions of a rational approach? These are some of the questions I ask myself when confronted with the love some intellectuals show for what they think are old things. To find an answer, I took a look at history and examined how early 'rational' critics of tradition proceeded and how their observations were received. More especially, I analysed what Xenophanes had to say about the traditions of his time.

Xenophanes was one of the first Western intellectuals. Like many of his successors he was a conceited bigmouth. Unlike them, he had considerable charm. He did not present well constructed arguments – this is why Aristotle called him 'somewhat uncouth' (*agroikoteros*: *Met*. 986b27) and advised his readers to forget about him – but effective one-liners. He crisscrossed Greece and Ionia singing the old stories, but he also criticised them and made fun of them. 'He dared, he, a Greek of the sixth century, to reject traditional tales as old inventions!' writes Hermann Fränkel (*Wege und Formen Frühgriechischen Denkens*, Munich 1968, p. 341). He still used the old forms such as the epic form and elegies. The fragment Fränkel refers to (fragment B 1 in the numbering of Diels-Kranz) sounds about as follows:

Clean is the floor, clean are the hands and the cups; and the garlands freshly now woven, are put on the heads by the boy.
Redolent balsam preserved in the phial is brought by another, exquisite pleasure lies waiting for us in the bowl;

and a different wine, with the promise of never bringing displeasure,
soft tasting and sweet to the smell, stands here in the jar.
And in the centre the incense dispenses the holy perfume;
cool water is there, full of sweetness and clear to the eye.
Behold the goldyellow loaves and, on the magnificent tables,
overflowing abundance of cheese and rich honey.
And in the centre an altar fully covered with flowers
and festive songs sounding all over the house.
But first it is proper for well disposed men to the god to pay tribute
with words which are pure and stories that fit the occasion;
then, after the common libations and the prayer for strength
                              to act wisely
(the most important concern, preceding all others)
it is not hybris to fill the body with drink – provided
only the old ones need later a slave to get home.
And I praise the man who, having imbibed, can still remember
how much he achieved and how he followed the virtues.
Let him not tell us of battles conducted by Titans and Giants
or even Centaurs – the fantasies of our fathers;
or of civic dissension – not useful are these events.
But one should always pay respect to the gods.

This poem has various interesting features. First, the surroundings: it is a somewhat restrained party where one thinks of the
gods and does not drink to excess. While some poets, like
Alkaeus, praised drinking for its own sake, and while those who
imitated the Lydians 'were so corrupted that some of them, being
drunk, saw neither the rising nor the setting of the sun' (Athenaeus's paraphrase of the end of fragment 3), Xenophanes advises
his drinking companions to drink with moderation so that only
the elderly will need a slave to get home. We owe the fragment to
precisely these observations: the physician Athenaeus of Attalea
who lived in the first century B.C. noticed its relevance for
medicine and quoted it in his book on dietetics.

A second interesting feature is the content of the conversations. They are not about wars or epical topics; they are about
the personal experiences of the participants – 'how much they
achieved and how they followed the virtues'. According to Xenophanes, these matters are furthered neither by Homer (who even
in democratic Athens was the basis of formal education: see
T.B.L. Webster, *Athenian Culture and Society*, University of
California Press 1973, chapter 3) nor by the modern craze for
athletics:

Let him be swift on his feet and in this way defeat all the others;
let him excel in five ways in the grove of the god
here in Olympia, close to Pisanian waters; let him
wrestle, or master the painful profession of boxing
or the terrible contest, known to all as Pankration –
greater would be his honour in the eyes of his neighbours.
Excellent seats would be given to him at the fights and the games
he could eat what he wanted and eat it at public expense;
in gifts they would drown him and permanent property would
. be his due
and this also if he had proved his ways with a horse,
he, who is lower than I. For my wisdom is better by far
than the brute power of men and of swiftfooted horses.
No, the custom that puts rugged strength over useful achievements
is without sense and should not be further encouraged.
Small is the gain for the city that harbours an excellent boxer
or a fivefold contestant or winner in wrestling,
or an excellent runner who, among all professions
engaged in competing, is by far the most praised.
Short is the pleasure the city derives from a contest in Pisa
for it does not fill the stores of the town.

'The greedy way in which these men (the athletes) ate, does not surprise us', wrote Athenaeus who also preserved this fragment. 'All participants in the games were invited to eat a lot and also to exercise a lot.' To set them up as examples and to revere them is of no use to the city, says Xenophanes.

However, Xenophanes did not only oppose the cultural tendencies of his time. According to the opinion of most modern thinkers, he also revealed their foundations and *criticized* them. Above all he criticized the idea that there exist gods who resemble humans, who are cruel, angry and treacherous like the heroes of the epic and who influence history. The critique, his later admirers say, led to the rise of rationalism. Is this true? Are Xenophanes' objections to traditional forms of thought really as penetrating and as fertile as is believed by many philosophers? Do they really force us to abandon the old idea of gods who have human features and act in this world?

Xenophanes's 'argument', as we know it, is very brief. It consists of the following comments:

Everything humans despise and condemn and try to avoid,
theft, and adultery, and lying deception of others
Homer and Hesiod respectfully brought to the gods . . .

[frgg 11,12]

But the mortals consider that gods were created by birth
that they wore clothes, had voices, and also a form.
But if cattle, or lions, or horses had hands, just like humans;
if they could paint with their hands, and draw and thus create
pictures –
then the horses in drawing their gods would draw horses; and cattle
would give us pictures and statues of cattle; and therefore
each would picture the gods to resemble their own constitution.

Aethiopian gods – snubnosed and black
Thracians – blue-eyed and blond (incomplete)
[frgg. 14, 15, 16]

Here is what some modern writers have said about these lines. Guthrie, *A History of Greek Philosophy*, Vol. i, Cambridge 1962, p. 370 speaks of 'destructive criticism'. Mircea Eliade, otherwise a very clever gentleman, praises 'Xenophanes' acute criticism' (*Geschichte der Religiösen Ideen*, Vol. 2, Herder 1979, p. 407). And Karl Popper, who has dragged Xenophanes up and down the countryside as one of his most important predecessors, reads the fragments as the 'discovery that the Greek stories about the gods cannot be taken seriously because they represent the gods as human beings' (*Auf der Suche nach einer besseren Welt*, Munich 1984, p. 218); he, too, speaks of a 'criticism'.

Xenophanes' positive views about god, or his 'theology', are contained in the following lines:

One god alone is the greatest, the greatest of gods and of men
not resembling the mortals, neither in shape nor in insight.

Always without any movement he remains in a single location
as it would be unseemly to walk now to this, now to that place.

Totally vision, totally knowledge, totally hearing.

But without effort, by insight alone, he moves all that is.
[frgg. 23, 26, 24, 25]

It is interesting to trace the effect this doctrine had in antiquity. We have quotations of key phrases in Aeschylus (cf. the first appendix to Guido Calogero, *Studien über den Eleatismus*, Darmstadt 1970) and we have a comment by Timon of Phleios, a pupil of Pyrrho the sceptic (quoted in Diogenes Laertius and, with a slight difference, in Sextus Empiricus, *Hypot.*, 224 – A 35 in Diels/Kranz). Timon writes:

Xenophanes, semi-pretentious, made mincemeat of Homer's decep-
tions,
fashioned a god, far from human, equal in all his relations,
lacking in pain and in motion, and better at thinking than thought.

'Far from human', Timon calls the gods of Xenophanes – and he
(it?) is indeed inhuman, not in the sense that anthropomorphism
has been left behind but in the entirely different sense that
certain *human* properties, such as Thought, or Vision, or
Hearing, or Planning, are monstrously increased while other,
balancing features such as tolerance, or sympathy, or pain have
been removed. 'Always, without any movement, he remains in a
single location' – like a king, or a high dignitary for whom 'it
would be unseemly to walk now to this, now to that place'. What
we have is not a being that transcends humanity (and should
therefore be admired?) but a *monster* considerably more terrible
than the slightly immoral Homeric gods could ever aspire to be.
These one could still understand; one could speak to them, try to
influence them, one could even cheat them here and there, one
could prevent undesirable actions on their part by means of
prayers, offerings, arguments. There existed personal relations
between the Homeric gods and the world they guided (and often
disturbed). The God of Xenophanes *who still has human
features*, but enlarged in a grotesque manner, does not permit
such relations. It is strange and, at least to me, somewhat fright-
ening to see with what enthusiasm many intellectuals embrace
this monster, regarding it as a first step towards a 'more sublime'
interpretation of divinity. On the other hand, the attitude is also
very understandable for the remaining human features are
features many intellectuals would love to possess: pure thought
made efficient by the power of moving everything from afar,
supervision, superhearing (for picking up intellectual gossip?) –
and no feelings.

I summarise: Xenophanes mocks the traditional gods because
of their anthropomorphic features. What he offers in their place
is a creature that is still anthropomorphic, but inhuman, adding,
incidentally (frg. 34), that he has no idea of what he is talking
about ('what I said about the gods no human has ever seen nor
will he ever know it'). And *this* Popper calls the '*discovery* that
the Greek stories about the gods cannot be taken seriously
because they represent the gods as human beings'.

I now turn to the critical fragments and my question is: are we
here dealing with a *criticism* or simply with a *rejection* of the idea

of regional gods that share the properties of the region they dominate? The answer is: the latter. The rejection becomes a criticism if we can assume

(A) that the concept of a god (or, speaking more generally, the concept of a Truth or of a Being) that changes from one culture to the next is not valid anywhere or, conversely, that a fitting concept of a deity (or a fitting concept of truth or being) must be valid everywhere and

(B) that the receiver of the criticism accepts (A), at least implicitly. Only then will the mockery hit its aim. Otherwise the opponent can always say: 'you are not speaking of our gods who are tribal gods, take care of us, look like us, live in accordance with our customs but have superhuman powers. What you are speaking of is an intellectual monster of your own invention which you use as a measure for all other gods. But this has nothing to do with us.' The mockery can even be inverted, as Timon's characterization shows: 'You, Xenophanes,' such inverse mockery could point out, 'are jealous of Homer's fame, so you want to outdo him and invent a god of your own, bigger than all the other gods, more stiff in his behaviour and more intelligent even than you.'

Many modern authors praise Xenophanes for making assumption (A). Not all of them are sincere in their praise, for not all of them believe that the world is ordered by divine powers. What these writers have in mind is not a superperson but something more abstract such as a law of nature, or a universal truth, or a uniform material. Disregarding this feature of Xenophanes's popularity, we must still point out that not all people accepted proposition (A) and that there existed authors and entire cultures, *both before and after Xenophanes*, who explicitly denied it. Thus Poseidon says in *Ilias* 15, 187ff [Lattimore tr.]:

. . . Since we are three brothers born by Rheia to Kronos,
Zeus, and I, and the third is Hades, lord of the dead men.
All was divided among us three ways, each given his domain.
I when the lots were shaken drew the grey sea to live in
forever; Hades drew the lot of the mists and the darkness,
and Zeus was allotted the wide sky, in the cloud and the bright air.
But earth and high Olympos are common to all three. Therefore
I am no part of the mind of Zeus. Let him in tranquillity
and powerful as he is stay satisfied with his third share.

According to this passage the natural world, just like the political world, is subdivided into regions which are subjected to different (natural) laws. E.M. Cornford commented on the passage and explained the terms it contains (*From Religion to Philosophy*, New York 1965, p. 16). *Moira*, translated as 'share', means 'part', 'allotted part' – this is also the original sense of 'fate' or 'destiny'. Poseidon's objections show that gods like humans have their *moirai*: each god is given a well-defined part of the world as his field of action. The parts are not only separated from each other, they are also qualitatively different (sky, water, darkness) and adumbrate the elements, which started as regions with certain qualities attached to them and only later became substances that could wander around in the cosmos. The region allotted to a god also determines his status (*time*) – it determines his position in a quasi-social system. The status is occasionally called his privilege (*geras*). Inside the region the rule of the god is not questioned; but he must not transgress its boundaries or else he encounters offended resistance (*nemesis*). Thus the world at large is seen as an aggregate with different divinities ruling over different parts of it: (B) is not correct.

The aggregate character of the Homeric world was not restricted to the very large – it is found in its smallest parts. There are no *concepts* that forge the human body and the human soul into a unit, there are no *means of representation* that would enable the artists to give optical expression to such a unity. Both conceptually and optically human beings are like rag dolls, sowed together from relatively isolated elements (upper arm, lower arm, trunk, neck, head with an eye that is simply put into its region without 'looking' at anything) and functioning as transit stations for events (ideas, dreams, feelings) that may arise elsewhere and only briefly merge with a particular human being. Action in our sense does not exist in this world; a hero does not decide to bring about a certain event and then cause it, he *finds himself involved in* one series of actions rather than in another and his life *develops* accordingly. All things, animals, carriages, cities, geographical regions, historical sequences, entire tribes are presented in this 'additive' manner – they are aggregates without 'essence' or 'substance'.

The same is true of world views. In religion we have an opportunistic eclecticism that does not hesitate to add foreign gods to those already accepted, provided their presence promises some advantage; different versions of the same story survive side by side (this was raised to a principle by Herodotus – vii, 152,3: *legein ta legomena*) and even the 'modern' and already rather desiccated ideas of the Ionian philosophers (Thales,

Anaximander, Anaximenes) do not combat tradition. There is no coherent *knowledge*, i.e. no uniform account of the world and the events in it. There is no comprehensive *truth* that goes beyond an enumeration of details, but there are many *pieces of information*, obtained in different ways from different sources and collected for the benefit of the curious. The best way of presenting such knowledge is the *list* – and the oldest scientific works were indeed lists of facts, parts, coincidences, problems in several specialised domains. The gods have complete knowledge. This does not mean that their glance penetrates the surface and perceives a hidden unity beneath events – they are not theoretical physicists or biologists – but that they have the most complete lists at their disposal. Even the early notions of validity agree with this situation: *nomos* comes from *nemein*; in the *Iliad* this word has the sense of distributing, or attributing to a certain region. (Details and further citations are given in chapter 17 of my book *Against Method*, London 1975, and in my forthcoming book *Stereotypes of Reality*.)

To sum up: (A) and (B) are not applicable to the Homeric world; Xenophanes *rejects* this world view, but he has not given us any *arguments* against it.

Turning now to authors who wrote after Xenophanes, we notice that some of the most intelligent writers either disregarded him, or went a different way. Aeschylus, who was 'strongly influenced' by Xenophanes (Calogero, *op. cit.,* p. 293, footnote 16), on the one hand gave the gods greater and more spiritual power and thereby made them less human; on the other hand he let them participate in the activities of the city (Athene, in the last part of the *Oresteia*, presides over a council containing Athenian citizens and casts her vote side by side with them) and so brought them closer to human concerns. The gods of Aeschylus also acted less arbitrarily and more responsibly than the Homeric gods, arbitrariness and responsibility being again measured by the standards of the city; this brought them closer even to human *modes of action* than were the gods of Homer and Hesiod. And, of course, the gods of Aeschylus were still the old gods, there were many of them, not just the one Xenophanean power-monster.

Sophocles then revived the arbitrariness of the Homeric gods. Trying to explain the seemingly irrational way in which good and bad fortune is distributed among humans, he attributed it to the actions of equally wilful and irrational gods (see for example

Electra, 558ff). Herodotus, whose sentence construction (*lexis eiromene*) and whose tolerance towards conflicting versions of the same story already formally reflects the aggregate view, supported the existence of divine influences with empirical arguments. His analysis of social laws and customs employed a regional notion of validity. It may be summarised in the following statement:

> Customs, laws, religious beliefs rule, like kings, in restricted domains. Their rule rests on a twofold authority – on their *power* (which is the power of those who believe in them) and on the fact that it is *rightful power*.

Protagoras extended this view from laws and customs to all matters of concern to humans, 'ontological matters' included. This was the *relativism of Protagoras*.

The relativism of Protagoras denies the two assumptions that turn Xenophanes' mockery into an argument, and thus agrees with basic principles of the Homeric world view. This shows that the followers of Protagoras and Herodotus were not lazy bums who, having reached the limits of their city, their nation or their cosmic region, stopped searching for a universal Truth or universal Canons of Validity and remained content with a collection of local opinions. Their philosophy was an exact mirror image of a world that had been inhabited by their forefathers and that still guided the thoughts and the perceptions of their contemporaries. Why else would the Platonic Socrates constantly run into people who answered questions such as 'What is knowledge?' or 'What is virtue?' or 'What is courage?' by giving *lists*? (cf. on this point Dover, *Greek Popular Morality*, Berkeley and Los Angeles 1978). But if the world is an aggregate of relatively independent regions, then any assumption of universal laws is *false* and a demand for universal norms *tyrannical*: only brute force (or seductive deception) can then bend the different moralities so that they fit the prescriptions of a single ethical system. And indeed, the *idea* of universal laws of nature and society arose in connection with a life-and-death battle: the battle that gave Zeus the power over the Titans and all other gods and thus turned *his* laws into *the* laws of the universe (Hesiod, *Theog.*, 644ff).

The idea of a universal truth and a universal morality has played an important role in the history of Western thought (and Western political action). It is often regarded as a measure by which theoretical suggestions and practical achievements must be judged and it gave respectability to the relentless expansion of civilization into all corners of the world. The expansion in a very ironical way revealed the violent origin of the expanding cultures: Western achievements were only rarely asked for and Western colonizers were hardly ever invited to bless the Primitives (or the Chinese, or the Japanese, or the Indians) with their advanced ideas and their sublime habits. The comment that relativistic philosophies are of no importance when compared with this almost inevitable development is of course true but only confirms the (relativistic) idea that the popularity of philosophical positions is a result of power (or deception), not of argument: *the regionalism of natural phenomena was never overcome, neither by philosophers nor by scientists, while the regionalism of social phenomena was repressed or destroyed by violence, not shown to be inadequate by ethical reasoning.*

Regarding the first part of this statement, we have to realise that a unified view of the physical world simply does not exist. We have theories that work in restricted regions, we have purely formal attempts to condense them into a single formula, we have lots of unfounded claims (such as the claim that all of chemistry can be reduced to physics), phenomena that do not fit into the accepted framework are suppressed; in physics, which many scientists regard as the one really basic science, we have now at least three different points of view (relativity, dealing with the very large, quantum theory for an intermediate domain and various particle models for the very small) without a promise of conceptual (and not only formal) unification; perceptions are outside the material universe (the mind-body problem is still unsolved) – from the very beginning the salesmen of a universal truth cheated people into admissions instead of clearly arguing for their philosophy. And let us not forget that it was they and not the representatives of the traditions they attacked who introduced argument as the one and only universal arbitrator. *They* praised argument – *they* constantly violated its principles. Xenophanes' mockery is the first, the shortest and the clearest example of such doubletalk.

In the social domain the situation is even worse. Here we have not only a failure of theories, but of human decency as well. Only few proponents of intellectual and industrial progress regarded the great variety of views and cultures that populated the earth as a problem and hardly any politician or colonizer or developer was prepared to argue for things he could get by force (there are exceptions – but they are rare). Thus the increasing uniformity of 'civilised' societies does not show that relativism has failed; it only shows that power can eliminate all distinctions.

I conclude with an example which shows how uncritically the basic assumptions of Xenophanes (assumptions A and B) are accepted by some of our contemporaries. The example is not a theory, or a philosophical point of view, but a poem by Czeslaw Milosz. The poem is rather naive and its faults can be shown in a few lines. Does this mean that people have ceased thinking about matters that affect their lives and that empty phrases, artfully put together, are more powerful even than commonsense? Xenophanes had strange views – but he showed signs of intelligence (and humour). I cannot detect any such sign in the following:

INCANTATION

1   Human reason is beautiful and invincible;
No bars, no barbed wire, no pulping of books,
no sentence of banishment can prevail against it.
It establishes universal ideas in language
5   and guides our hand to write Truth and Justice
with capital letters, lie and oppression with small.
It puts what should be above things as they are
it is the enemy of despair and a friend of hope.
It does not know the Jew from Greek or slave from master
10   giving us the estate of the world to manage.
It saves austere and transparent phrases
from filthy discord of tortured words.
It says that everything is new under the sun,
opens the congealed fist of the past.
15   Beautiful and very young are Philo-Sophia
and poetry, her ally in the service of the Good.
As late as yesterday Nature celebrated her birth.
The news was brought to the mountains by a unicorn and an echo.
Their friendship will be glorious, their time has no limits.
20   Their enemies have delivered themselves to destruction.

'Destruction' (20) threatens the opponents of a non-regional Reason intent on 'manag[ing] the estate of the world' (10) without any 'filthy discord of tortured words' (12), i.e. without democratic discussion. This is true, but not in the sense intended by Milosz: 'destruction' did indeed remove all those small and well adapted societies that were in the way of the expansion of Western civilization, even though they tried to defend their rights with 'tortured words'. Noble Reason, on the other hand, is hardly 'invincible' (1); prophets, salesmen, politicians trample it underfoot, the alleged friends of reason distort it to make it fit their intentions. The sciences of the past have showered us with useful and terrible gifts – but without employing a single un-changeable and 'invincible' agency. The sciences of today are business enterprises run on business principles. Research in large institutes is not guided by Truth and Reason but by the most rewarding fashion, and the great minds of today increasingly turn to where the money is — which means military matters. Not 'Truth' is taught at our universities, but the opinion of influential schools. Not Reason, or Enlightenment, but a firm faith (in the Bible, or in Marxism) was the strongest preserving force in Hitler's prisons, as Jean Améry discovered. 'Truth', written 'in capital letters' (6), is an orphan in this world, without power and influence, *and fortunately so* for the creature Milosz praises under this name could only lead to the most abject slavery. It cannot stand diverging opinions – it calls them 'lies' (6); it puts itself 'above' (7) the real lives of human beings, demanding, in a way characteristic of all totalitarian ideologies, the right to rebuild the world from the height of the 'what should be' (7), i.e. in accordance with its own 'invincible' precepts (1). It refuses to recognize the many ideas, actions, feelings, laws, institutions, racial features which separate one nation (culture, civilization) from another and which alone give us *people*, i.e. creatures with *faces* (9).

This is the attitude that destroyed Indian cultural achieve-ments in the USA without so much as a glance in their direction, this is the attitude that is now destroying non-Western cultures under the guise of 'development'. Conceited and self satisfied is this faith in Truth and Reason for which a democratic discussion is but a 'filthy discord of tortured words' (12) – and also very uninformed: philosophy was never the 'ally' (15) of poetry, not in antiquity when Plato spoke of the 'ancient battle between philo-sophy and poetry' (*Rep.* 607b6f), not today when Truth is sought in the sciences and poetry is reduced to the expression of feel-ings. The reason of ordinary people trying to create a better and safer world for themselves and their children (which is reason with a small 'r' and not Reason written 'in capital letters': 6) has very little in common with these ignorant and irrational dreams of domination. Unfortunately commonsense is too common an instrument to impress intellectuals and so they abandoned it long ago, replaced it by their own conceptions and tried to redirect political power accordingly. We must restrict their influence, remove them from positions of power and turn them from the *masters* of free citizens into their most obedient *servants*.

3

Knowledge and the Role of Theories

1   Existence

The world we live in contains an abundance of things, events, processes. There are trees, dogs, sunrises; there are clouds, thunderstorms, divorces; there is justice, beauty, love; there are the lives of people, gods, cities, of the entire universe. It is impossible to enumerate and to describe in detail all the incidents that happen to an individual in the course of a single boring day.

Not everybody lives in the same world. The events that surround a forest ranger differ from the events that surround a city dweller lost in a wood. They are different events, not just different appearances of the same events. The differences become evident when we move to an alien culture or a distant historical period. The Greek gods were a living presence; 'they were there'.[1] Today they are nowhere to be found. 'These people are farmers', E. Smith-Bowen writes about an African tribe she visited:[2] 'to them plants are as important and familiar as people. I'd never been on a farm and I am not even sure which are begonias, dahlias, or petunias. Plants, like algebra, have a habit of looking alike and being different, or looking different and being alike; consequently mathematics and botany confuse me. For the first time in my life I found myself in a community where ten-year-old children weren't my mathematical superiors. I also found myself in a place where every plant, wild or cultivated, had

1. Ulrich von Wilamowitz-Möllendorf, *Der Glaube der Hellenen* I, Darmstadt 1955, p. 17. Cf. also W.F. Otto, *Die Götter Griechenlands*, Frankfurt 1970.
2. E. Smith-Bowen, *Return to Laughter*, London 1954, p. 19.

a name and use and where every man, woman and child literally knew hundreds of plants . . . [my instructor] simply could not realise that it was not the words but the plants that baffled me.'

Bafflement increases when the objects encountered by the explorers are not just unfamiliar, but inaccessible to their ways of thinking. Inflectional languages posit things having properties and standing in certain relations to each other: snow whirls around in the wind, it lies on the ground, it rises like a curtain in a storm. The same thing, snow, gets involved in a variety of episodes. The Delaware Indians, on the other hand, approach the world like painters who use a different brush, different colours and a different type of stroke for each snow episode.[3] They not only fail to notice 'snow', they cannot even imagine that 'it' exists. 'Where we are likely to use completed, fully developed words or parts of speech, the Eskimo creates new combinations, invented especially for the purpose to meet the challenge of each and every situation. Concerning the formation of words, the Eskimo is constantly *in statu nascendi* . . . words are born on his tongue under the impact of the moment.' Simple conversations employ 10,000 to 15,000 particles. Speech is poetry and poetry is common – it is not the exclusive possession of specially gifted and separately trained individuals. Space, time, reality change when we move from one language to another. According to the Nuer, time does not limit human action, it is part of it and follows its rythm: '. . . the Nuer . . . cannot speak of time as though it was something actual, which passes, can be waited for, can be saved and so forth. I do not think that they ever experience the same feeling of fighting against time, or of having to co-ordinate activities with an abstract passage of time, because their points of reference are mainly the activities themselves, which are generally of a leisurely character . . .'[4] For the Hopi a distant event is real only when it is past, for a Western businessman it occurs in the presence of events he is participating in. The worlds in which cultures unfold not only contain different events, they also contain them in different ways.

3. For further examples of the way in which 'agglutinating' (Humbold) or 'polysynthetic' (Duponceau) languages represent reality cf. Werner Müller, *Indianische Welterfahrung*, Stuttgart 1976, with literature. The quotation is from page 21. Useful comments on linguistic classifications are found in chapter 7 of Yuen Ren Chao, *Language and Symbolic Systems*, Cambridge 1968.

4. E.E. Evans-Pritchard, *The Nuer*, Oxford 1940, p. 103. For the Hopi cf. B.L. Whorf, *Language, Thought and Reality*, MIT Press 1956, p. 63.

2 Knowledge

Living in a particular world, an individual needs knowledge. An enormous amount of knowledge resides in the ability to notice and to interpret phenomena such as clouds, the appearance of the horizon on an ocean voyage,[5] the sound patterns in a wood, the behaviour of a person believed to be sick – and so on. The survival of individuals, tribes, and entire civilizations depends on this kind of knowledge. Our lives would fall apart if we could not read people's faces, understand their gestures, react correctly to their moods.

Only a fraction of this 'tacit' knowledge [6] can be articulated in speech and if it is, then knowledge of the same kind is needed to connect the words with the corresponding actions. Knowledge is contained in the ability to perform special tasks. A dancer has knowledge in her limbs, an experimentalist in hands and eyes, a singer in the tongue, the throat, the diaphragm. Knowledge resides in the ways we speak, the flexibility inherent in linguistic behaviour included:[7] linguistic knowledge is not stable, it contains elements (ambiguities, analogies, patterns of analogical reasoning) that can undermine any particular stage of it.

Language and perception interact. Every description of observable events has what one might call an 'objective' side – we recognize that it 'fits' a particular situation – and 'subjective' ingredients: the process of fitting description to situation modifies the situation. Features lacking in the description tend to recede into the background, outlines emphazised by the description become more distinct. The changes are noticed when the description is first introduced; they disappear when using it has become routine. The apparent objectivity of familiar 'facts' is a

5. J.G. Herder, 'Journal meiner Reise im Jahre 1769', *Sämtliche Werke*, Vol. iv, Berlin 1878, esp. pp. 356f, commented on the way in which individual observations were systematised by sailors, giving rise to interesting combinations of empiricism and superstition. For the 'preoccupation with exhaustive observation and the systematic catologuing of relations and connections' shown by indigenous peoples cf. C. Lévi-Strauss, *The Savage Mind*, Chicago 1966. The quotation is from page 10.

6. Cf. M. Polanyi, *The Tacit Dimension*, Garden City, New York, 1966. For the background cf. the same author's *Personal Knowledge*, London 1958. Oliver Sacks, *The Man Who Mistook his Wife for a Hat*, New York 1987 (first published in 1970), contains case studies showing what happens when tacit knowledge of a certain kind is no longer available.

7. On this point cf. Chapter 10.

result of training combined with forgetfulness and supported by genetic dispositions; it is not the result of deepened insight. [8]

What is true of languages is true of all means of representation. A caricature has an 'objective' core – this is how we recognize its target – but it also invites us to look at the target 'subjectively', through the eyes of a special group, or of an individual with a special vision (for example, Kokoschka's portraits). The new way of looking may intrude to such an extent that recognition becomes impossible without it – it is now 'part of reality' or, to turn the argument around, the original 'reality' was but another 'subjective', but popular view. Novels, fables (with or without an explicit moral), tragedies, poems, liturgical events such as Holy Mass, conceptual considerations, scientific arguments, scholarly histories, newscasts, documentaries initiate, or reinforce, or give content to, similar developments: events are structured and arranged in special ways, the structures and the arrangements gain in popularity, they become routine, intellectuals interested in perpetuating the routine provide it with a 'foundation' by showing that and how it leads to important results (most theories of knowledge are longwinded defences of existing or incipient routines). Far-reaching practices and views have been supported by a 'reality' that was shaped by them in the first place.

The transformations are most noticeable in history, politics and the social sciences. A social history of the French Revolution shares only names with narratives that concentrate on kings, generals, wars. 'The images which authors hand over to later generations are formalised and occasionally completely contrary to truth . . . There is sometimes talk about the "Storming of the

---

8. Instructive cases are found in the studies of psychologists with a phenomenological orientation, for example in D. Katz's pathbreaking essay *Die Erscheinungsweise der Farben, Zeitschrift für Psychologie und Physiologie der Sinneorgane*, Ergänzungsband 7, Leipzig 1911. The essay explores the qualities of colours as perceived by an average observer and introduces the distinction between spectral colours (such as the colour of the sky) which seem to have depth and surface colours (the colour of an apple) which are restricted to a clearly defined surface. The study was not without difficulties. Many observers produced vague and practically useless reports. 'Observers must be prepared', says Katz (p. 41). Observers were prepared by receiving the 'correct' description (or an approximation of it), but camouflaged as a question. The question evoked the phenomenon which then remained tied to the description: the resulting 'objectivity' had 'subjective' causes. Transitions such as these occur in all subjects that are tied to observation and they define reality for them. For the arts cf. A. Ehrenzweig, *The Hidden Order of Art*, Berkeley and Los Angeles 1967; relevant episodes in the history of the sciences are discussed by W. Lepenies, *Das Ende der Naturgeschichte*, Munich and Vienna 1976.

Bastille" though in reality the Bastille was not stormed by anyone
– the 11th. of July 1789 was merely an episode in the French
Revolution; the people of Paris entered the prison without difficulty and found there only few prisoners. But just this capture
of the Bastille became the national holiday of the revolution.' So
writes I. Ehrenburg; and S.E. Luria notes that 'being present at a
major social dislocation is a peculiar experience. To a historian
such an event represents a nodal occurrence of causes and
effects; to a journalist a mosaic of vignettes with human interest.
In the hands of a great novelist – the plague of Milan for Manzoni
or the retreat from Moscow for Tolstoy – the great dislocations of
human life become inspiration for revealing the human condition
at its worst and its best. But for the individual participant who is
not in any of these literary classes the great event translates itself
into a composite of minor occurrences, each of which amounts to
problem solving – making do.' [9] There is no better way of describing how 'literary classes' create 'significant historical developments' out of chaotic (and, for them, uninteresting) events in
the lives of (for them) undistinguished people and how non-
entities become 'major historical figures' in the process.

The conflict between the 'realities' that emerge from different
approaches becomes accentuated when one of the approaches is
part of a popular political (or scientific, or religious) movement.
An example is the recent debate about the Warsaw Ghetto
Uprisings. Participants and commentators of a leftist or a nation-
alist outlook regarded them as outbursts of heroic proportions.
For Dr. Marek Edelman, who also participated but who did not
share these ideologies, the uprisings were insignificant fluctuations in an absurd sequence of events. Another example is the
trial of Galileo. It was a minor event in the historical context of
the time. Galileo had given a promise, had broken his promise
and had tried to hide behind lies. A compromise was sought, a
compromise was found. Galileo continued writing and continued

9. I. Ehrenburg, *People and Life*, Memoirs of 1891–1917, London 1961, p. 8.
Cf. also Ehrenburg's description of the situation in Leningrad during the Russian
October 'revolution'. For similar reports on events constituting the 'French
Revolution' cf. G. Pernoud and S. Plaissier, *The French Revolution*, New York
1960.

The second quotation is from S.E. Luria, *A Slot Machine, A Broken Test
Tube*, New York 1985, p. 26. The Warsaw uprisings are discussed in Norman
Davies, 'The Survivor's Voice', *The New York Review of Books*, Nov. 20, 1986,
pp. 21ff. Learned comments on the general situation are found in Stuart Hughes,
History as Art and Science, New York 1964.

smuggling contraband out of Italy. He was more fortunate, less determined and certainly less courageous than Bruno. But modern scientists, needing a hero and regarding scientific knowledge as inviolate as the Church once viewed the Host, turned the tribulations of an anxious crook into a clash of giants.

Knowledge can be stable and it can be in a state of flux. It may be available in the form of public beliefs shared by all, and it may reside in special individuals. It may reside in them in the form of general rules that are learned by rote, or as an ability to treat new situations in an imaginative way. The laws of Hammurabi and the Draconian code were written down, the Mosaic law was for a long time part of a common oral tradition that could be quoted whenever the need arose, the grammatical 'laws' of American-Indian languages and the judgements of 'the elders in session on benches of stone' (*Iliad* 18, 503f) are inventive reactions to particular problems. Even a modern judge who has to rely on written guidelines and volumes of past court decisions needs intuitive knowledge to render his verdict.

The invention of writing created new types of knowledge and started interesting debates. Early forms of writing had no relation to speech. They were aids in calculation and means for recording business transactions.[10] The use of writing for preserving information of a more substantial kind was criticized by Plato. 'You know, Phaedrus,' says Socrates (*Phaedrus*, 275d2ff),

> that's the strange thing about writing, which makes it truly analogous to painting. The painter's products stand before us as though they were alive, but if you question them, they maintain a most majestic silence. It is the same with written words; they seem to talk to you as though they were intelligent, but if you ask them anything about what they say, from a desire to be instructed, they go on telling you the same thing forever. And once a thing is put in writing, the composition, whatever it may be, drifts all over the place, getting into the hands not only of those who understand it, but equally of those who have no business with it; it doesn't know how to address the right people, and not address the wrong. And when it is ill treated and unfairly abused it always needs its parent to come to its help, being unable to defend or help itself.

10. Information in chapter 5 of D. Page, *History and the Homeric Iliad*, Berkeley and Los Angeles 1966, and in chapter 7 of J. Chadwick, *The Decipherment of Linear B*, Cambridge University Press 1958. A general survey is I.G. Gelb, *A Study of Writing*, University of Chicago Press 1963.

Gaining knowledge, according to this account, is a process that involves a teacher, a pupil and a (social) situation shared by both; the result, knowledge, can only be understood by those who participated in it. Written notes aid them in remembering the stages of their participation. Being incapable of replacing the process, they are useless for outsiders. Later on, when philosophy changed into an academic discipline and when treatises and research papers increased in importance, the historical components of knowledge receded into the background. Knowledge was defined as what can be extracted from a written page. [11] Today it is taken for granted that the history of a particular piece of information has no relevance for understanding its content. 'What matters in science', writes S.E. Luria (*op. cit.*, p. 123) in a passage already quoted and criticized,

> is the body of findings and generalizations available today: a time-defined cross section of the process of scientific discovery. I see the advance of science as self erasing in the sense that only those elements survive that have become part of the active body of knowledge. The model of the DNA molecule worked out by Crick and Watson stands on its own merits . . . The . . . story of how the DNA model was achieved, humanly fascinating as it may be, has little relevance for the operational content of science.

Most philosophers of science agree, and add their blessings by distinguishing between a context of discovery and a context of justification: the context of discovery tells the history of a particular piece of knowledge, the context of justification explains its content and the reasons for accepting it. Only the latter

---

11. Plato consciously chose the dialogue over drama, the epic, the scientific treatise, lyrical poetry and the instructive speech as a way of presenting his views. Traditional Greek education and research had utilised all these forms (there existed as yet no distinction between knowledge and emotion, the arts and the sciences, truth and beauty). Letters played an important role in the 17th and 18th centuries. Galileo described some of his most important ideas in carefully formulated letters to special individuals. The letters were copied and handed around. Father Mersenne was a central exchange station for letters concerning the philosophy of Descartes. Newton's early papers on the theory of colours were letters to Henry Oldenburg, the secretary of the Royal Society; the debate caused by them was conducted by letters to Newton, or from Newton, via Oldenburg. Popular treatises (Galileo's *Dialogo*, written as an exchange between three characters, and Newton's *Opticks*) introduced new discoveries to a wider public. The personal element was not missing in either case. Only few modern scientists and philosophers make conscious decisions about matters of style – and there is no need for them to make such decisions, for most scientific journals have well defined editorial policies.

context concerns the scientist (and the philosopher cleaning up behind him). Yet the Platonic conception of knowledge (as explained in the above quotation from the *Phaedrus*) is on the rise again. Large parts of modern mathematics, physics, molecular biology, geology rest on an oral culture that contains unpublished results, methods and conjectures, and gives meaning to what is already available in print. Workshops, conferences, seminar meetings in leading research centres do not merely add information to the content of textbooks and research papers, they explain this content and make it clear that it cannot stand on its own feet. Pure mathematics more than any other subject has become that 'living discourse' Plato regarded as the only true form of knowledge. The 'hermeneutic' school in philosophy, though far less clear and far more verbose than Plato, tries to show that even the most 'objective' written presentation is comprehended only by virtue of a process of instruction that conditions the reader to interpret standard phrases in standard ways and that would collapse without a community of thinkers arguing in this manner: there is no escape from history and personal contact, though there exist powerful mechanisms creating the illusion of such an escape.

3   Forms of Knowledge

Knowledge orders events. Different forms of knowledge engender different ordering schemes. *Lists* played an important role in the growth of Near Eastern (Sumerian, Babylonian, Assyrian, early Greek) knowledge.[12] Lists of words were collected by interpreters relating near Eastern languages to Akkadian, the common (diplomatic) language of the region. Assembling words under their proper determinative (a classifying sign of the cuneiform script), they achieved simple classifications of the corresponding things: an early form of science was created entirely for the convenience of translators. Lists of customs, rules, descriptions, individuals, and problems served lawgivers, navigators (lists of harbours along a route combined with coastal descriptions), travellers (itineraries), genealogists (whose lists of heroes and kings preceded more sophisticated forms of historical narrative), mathematical instructors (Babylonian lists of mathe-

---

12. W. von Soden, *Leistung und Grenzen Sumerisch Babylonischer Wissenschaft*, reprint Darmstadt 1965.

matical problems together with solutions and helpful hints). The frequent objections of early philosophers to *polymathi'e,* and the first answers Socrates receives to his inquiries about the nature of courage, wisdom, knowledge, and virtue, show that lists were not transitory pigeonholes but basic ingredients of Greek commonsense.[13]

Lists classify in one dimension. The classificatory schemes of botanists, zoologists, chemists (periodic system of elements) astronomers (the Russell-Hertzsprung diagram), physiognomists, elementary particle physicists and, so it seems, of many indigenous peoples[14] are multidimensional, but static. Temporal sequences are described in *stories.* Simple stories deal with simple processes in the lives of plants, animals, or human beings (medical histories, starting with the *Prognostics* of the Hippocratic Corpus). Complex stories leading up to the evocation of comprehensive cosmic patterns were used by archaic peoples to relate celestial changes (precession included) to gestation periods, wandering of herds, bird and fish migrations, the growth of vegetation, the phases of the moon and large scale social changes.[15] They preserved knowledge and initiated social events; they were theoretical astrosociobiology and social glue in one.[16]

Stories were used to explain features which later generations turned into abstract properties. The Homeric epic 'defined' basic social relations (such as the four cardinal virtues of archaic and classical Greece: courage, piety, justice and wisdom) by showing how they worked in concrete cases. Diomedes is courageous

13. For the role of lists of moral matters cf. K.J. Dover, *Greek Popular Morality at the time of Plato and Aristotle,* Berkeley and Los Angeles 1974. For details and further literature see chapter 17 of my *Against Method,* London 1975.

14. Cf. the work of Lévi-Strauss, beginning with *The Savage Mind,* Chicago 1966, and the numerous analyses of kinship relations. Lévi-Strauss emphazises the efficiency of primitive classificatory schemes and the fact that they go far beyond practical needs: 'The natives are also interested in plants which are of no direct use to them, because of their significant links with the animal and insect world', i.e. because they fit into comprehensive theoretical schemes. 'Even a child [of the Tyukyu archipelago] can frequently identify the kind of tree from which a tiny wood fragment has come and, furthermore, the sex of that tree, as defined by Kabirian notions of plant sex, by observing the appearance of its wood and bark, its smell, its hardness, and similar characteristics.' 'Several thousand Coahuila Indians never exhausted the natural resources of a desert region in South California, in which today only a handful of white families manage to subsist. They lived in a land of plenty, for in this apparently completely barren territory, they were familiar with no less than sixty kinds of edible plants and twenty eight others of narcotic, stimulant or medical properties' (op. cit., pp. 4f.).

(*Iliad* 5, 114f); his courage occasionally gets out of hand; he then behaves like a madman (330ff; 434ff). Not the author, but the listener (or, in our days, the reader) makes this judgement and derives the limits of courage from it. Wisdom receives a similar treatment. Odysseus often acts in a wise and well balanced manner. He is selected to speak to temperamental stars like Achilles, he is sent on difficult missions. But his wisdom, too, may change face and turn into slyness and deception (*Il.* 23, 726ff). The instances show what courage and wisdom are; they do not nail them down as does a logical definition.

Concepts so introduced are not abstract entities, they are not separated from things. They are aspects of things, on the same level as colour, swiftness, beauty of motion, expertise and the handling of weapons or of words. They are adapted to the circumstances in which they come to the fore and they change accordingly.[17] The diseases found in the empirical tracts of the Hippocratic Corpus are defined in the same way. They are not 'disease entities', i.e. abstract things and processes that can be separated from the suffering body, they are features of this body. They are discovered by inspection and presented in stories and they change as the body and the inspecting physician change.[18] Stories were used in the Middle Ages and then, again, in the

15. Details in Giorgio de Santillana and Herta von Dechend, *Hamlet's Mill*, Boston 1965 and A. Marshack, *Roots of Civilization*, New York 1972. Precession is discussed in de Santillana and von Dechend, pp. 56 ff. Opinions about the astronomical background of archaic knowledge (with literature) are examined in D.C. Heggie, *Megalithic Science*, London 1981. B.L. van der Waerden has succeeded in reconstructing a 'mathematical science which must have existed in the Neolithic Age between 3000 and 2500 B.C. and spread from Central Europe to Great Britain, to the Near East, to India and China:' see *Geometry and Algebra in Ancient Civilizations*, New York 1983, p. xi. Hans Peter Doerr, *Sedna*, Frankfurt 1984 discusses the way in which knowledge (artificially separated by Western observers) is intertwined with the rest of social life.

16. The combination has survived until today. Scientific subjects and special schools are held together by common beliefs and practices. A recent example is the history of the so-called phage group and its influence on (molecular) biology. Cf. J. Cairns, G.S. Stent and J.D. Watson, eds., *Phage and the Origins of Molecular Biology*, Cold Spring Harbor Laboratory, New York 1966 and Peter Fischer, *Licht und Leben*, Konstanz 1985, esp. p. 141, commenting on the group's unwillingness (or inability) to consider results formulated in different terms and obtained by different methods.

17. Early practical explanations of the virtues, their gradual replacement by theory, the renewed emphasis on practicality in Sophistic thought and the return of theory in Plato are discussed in Fritz Wehrli's excellent little book *Hauptrichtungen des Griechischen Denkens*, Stuttgart-Zurich 1964. Cf. also Bruno Snell, *The Discovery of Mind*, 1962, and my *Stereotypes of Reality*, forthcoming.

Enlightenment to instruct the Faithful and to teach them the subtle and apparently innocent ways in which the Devil (or Unreason) might insinuate themselves into their lives. The stories were sometimes written down and illustrated, on other occasions they were invented on the spot, acted out, and handed on by an oral tradition, but they could also be carefully prepared and exactly produced (Lessing's and Schiller's moralising dramas) or even chanted, as was the case with liturgical drama, an early predessor of opera.

Dramatic accounts are still more complex. They reveal and heighten features of our social lives that sound unproblematic when told in ordinary speech. Thus some ancient writers of tragedy showed that basic values were incoherent and that moral conflict was inevitable. The demonstration was concrete, the spectator was guided into the conflict and made to feel its power. Others tried to resolve conflicts in a manner compatible with existing political institutions. According to Aristotle tragedies, properly constructed, reveal universal laws of human existence and are in this respect 'more philosophical than history' (*Poetics* 1451a38b6f): social laws are buried under details; historians arrange the details according to their interests, or simply according to convenience, while the tragedian penetrates the layer of particular facts, finds what is universal and imposes it on the minds of the audience. He is researcher, social historian and PR man, all wrapped into one. Brecht demanded that what is universal 'be characterised in [its] historical relativity': the stage action was to present a real, but changeable process and it was to present it in a manner that showed the audience how change could be, or might have been achieved.[19]

Stories of the most diverse kinds – sketches, novelettes, fables, travel reports, descriptions of marvellous objects — enriched early Greek literature. Herodotus used all of them to adapt the narration of his *Histories* to the incidents and patterns he had encountered and wanted to describe. Some of his successors

18. For the difference between 'disease entities' and diseases as observable aspects of a person cf. O. Temkin, *The Double Face of Janus*, Johns Hopkins University Press, Baltimore 1977, chs. 8 and 30. Thomas Sydenham, who 'laid the foundations of clinical medicine' (K. Dewhurst, *Dr. Thomas Sydenham*, University of California Press 1966, p. 59) regarded it as a physician's duty to attend 'to the outer husk of things' (*Works*, ed. R.G. Latham, Vol. 2, 60) while John Locke who 'more than any other Englishman helped to bring Sydenham's ideas to the notice of foreign doctors' (Dewhurst, *op. cit.*,p, 56) explained how perception changes in the course of the development of knowledge: *Essay*, chapter 9.

avoided stories with a contemptuous shrug; obviously history was too sublime a subject to be presented in the same way as myths and fairytales. But stories are becoming popular again. To some modern writers they are the only form adapted to the complexities of human thought and action.[20] If we add sculpture, paintings, drawings, caricature, scientific illustrations or, in our own time, tapes, computer graphics, mathematical formulae, recordings, film, and holographic drama, we find an abundance of events, types of information and ordering principles, presenting knowledge of the most varied kind.

4   Philosophy and the 'Rise of Rationalism'

The social groups who prepared what is now known as Western rationalism and who laid the intellectual foundations for Western science refused to take this abundance at face value. They denied that the world was as rich and knowledge as complex as the crafts and the commonsense of their time seemed to imply. They distinguished between a 'real world' and a 'world of appearances'. As they presented the matter, the real world was simple, uniform, subject to stable universal laws and the same for all. New concepts (later called 'theoretical concepts') were needed to describe it and new subjects (epistemology and, later on, the philosophy of science) arose in the attempt to explain how it was related to the rest. Interestingly enough this 'unreal' rest was attributed to 'the many', that is, to common people and (unphilosophical) artisans.[21] From the very beginning, intellectuals claimed to possess insights unavailable to ordinary mortals.

19. Conflict pervades Sophocles's *Antigone*: cf. the analysis in chapter 3 of M.C. Nussbaum, *The Fragility of Goodness*, Cambridge University Press 1986. The emphasis on conflict was one of the reasons why Plato objected to tragedy and traditional religion: cf. his *Eutyphron*. For an attempt at a 'political' solution see the end of Aeschylus' *Eumenides*. Brecht explained his views in many places. The quotation is from his 'Kleines Organon für das Theater', section 36, *Gesammelte Werke*, Vol. 16, Frankfurt 1967.

20. Martin J.S. Rudwick, *The Great Devonian Controversy*, University of Chicago Press 1985, argues that even the development of (scientific) ideas is better served by narrative than by a 'conceptual account'. Using a well documented episode in the history of nineteenth-century geology, he shows that a scientific debate is too complex to be grasped by logicians. O. Sacks, *op. cit.*, p. 5 reports that according to R.A. Luria the effects of neural dysfunction 'would be best introduced by a story – a detailed case history of a man with a profound right hemisphere disturbance'. This is of course old hat for all clinicians, back to Hippocrates.

In developing their views, the 'philosophers' (a name soon applied to these groups) built, but they also destroyed. Like the invaders and conquerers before them, they wanted to transform the territory they entered. Unlike them they lacked physical power and therefore used words, not weapons, to achieve their aim. An enormous amount of their work (and of the work of scientists from Descartes and Galileo up to and including our own Nobel prizewinners) consisted in combating, ridiculing and, if possible, eliminating ideas and practices which, though well established, successful and advantageous for many people, did not conform to their idiosyncratic standards.

Almost all of them praised oneness (or, to use a better word, monotony) and denounced abundance. Xenophanes rejected the gods of tradition and introduced a single faceless god-monster. Heraclitus heaped scorn on *polymathi'e*, the rich and complex information that had been assembled by commonsense, artisans and his own philosophical predecessors, and insisted that 'what is Wise is One' (Diels-Kranz, B 40/41). Parmenides argued against change and qualitative difference and postulated a stable and indivisible block of Being as the foundation of all existence. Empedocles replaced traditional information about the nature of diseases by a short, useless but universal definition. Thucydides criticized Herodotus's stylistic pluralism and insisted on a uniform causal account. Plato opposed the political pluralism of democracy, rejected the view of tragedians such as Sophocles that (ethical) conflicts might be unresolvable by 'rational' means, criticized astronomers who tried to explore the heavens in an empirical way and suggested tying all subjects to a single theoretical basis. Entire armies of writers, teachers, and heads of schools led a 'longlasting battle' (Plato, *Republic* 607b6f) against traditional, undefined and fairly unregulated ways of thinking, speaking, acting, and arranging public and private lives.

In trying to evaluate the course of this battle, we must avoid confounding the interests of special groups with the fate of the world at large. European philosophy may have been a 'series of

21. For example, Parmenides rejected the 'ways of the humans' (Diels-Kranz, fragment B1,27), of 'the many' (B6,7) who, guided by 'habits based on much experience' (B7,3), 'drift along, deaf as well as blind, disturbed and undecided' (B6,6ff). Thus statements such as 'this is red' or 'that moves', which play an important role in the lives of ordinary human beings as well as of artists, physicians, generals, navigators, were summarily excluded from the domain of truth.

Philosophical attitudes towards 'the many' are discussed in Hans-Dieter Voigtländer, *Der Philosoph und die Vielen*, Wiesbaden 1980.

footnotes on Plato', as Whitehead remarked, but European history and European culture (which includes the castrato Farinelli, the Viennese comedy writer Nestroy, Hitler and my aunt Emma) certainly were not. Similarly, the 'battle', while perhaps filling the minds (and the papyrus rolls) of a few specialists, left the majority of people unmoved. In some areas there was not even a 'battle'. Xenophanes's god-monster left traces in Aeschylus, but not in Sophocles whose gods can again act out of pure spite: history, for Sophocles, was too irrational to have been created by rational gods. (Herodotus seems to have shared this view.) Popular religion remained unchanged.

The philosophers also failed to make the use of theoretical concepts a popular habit. The answers the Platonic Socrates received to his inquiries show that while people were prepared to grant uniformity to numbers (*Theaetetus*, 148b6ff) and bees (*Meno*, 72b6f), they balked at extending it to complex social relations such as knowledge or virtue.[22] Parmenides initiated some technical debates (Zeno, Plato's *Parmenides* and *Sophist*), but the public effect of his ideas was more in the realm of comedy than of knowledge (cf. Plato's *Euthydemus* whose sophisms come from Parmenides' identification of thought and being in B2,5ff). Even Parmenides's followers found his principles too extreme to be acceptable.[23] Opponents rejected the entire approach. 'Although these opinions appear to follow if one looks at the arguments,' wrote Aristotle,[24] 'still to believe them *seems*

22. As Plato presents the matter, Theaetetus and Meno seem to have made a simple mistake: Socrates asked for one thing, knowledge here, virtue there, and got many. But this criticism counts as an objection only if one *word* always means exactly one *thing* – the point at issue. It also assumes that there existed only one word for 'knowledge'; but the short passage *Theaet.* 145d4–e6 already contains four different epistemic terms. Finally, Meno's [Theaetetus'] reasoned resistance ('I don't think I can do with virtue [knowledge] what seems obvious for bees [numbers]') indicates that we are dealing with more than a sophomoric error. And we are indeed: what is at stake is an entire tradition of concept formation and concept explication.

23. The atomists, Empedocles and Anaxagoras accepted Parmenides' notion of Being but tried to retain change. To do this they introduced a (finite or infinite) number of things, each possessing either some or all Parmenidean properties: the atoms of Leucippus and Democritus were indivisible and permanent, but infinite in number, the elements of Empedocles finite in number, permanent, divisible into regions, but not divisible into further substances (the four elements of Empedocles, the Hot, the Cold, the Dry and the Moist were therefore different from any known substance), while Anaxagoras assumed the permanence of all substances.

to stand next door to lunacy when one considers practice. For in fact no lunatic seems to stand so far outside as to suppose that fire and ice are one' (an assumption implicit in the Ionian philosophy of nature and made explicit by Parmenides). Artisans who expressed their opinions in writing had raised similar objections long before (see *Ancient Medicine*, chapters 15 and 20). Stonemasons, metalworkers, painters, architects, and engineers apparently remained mute, but they left buildings, tunnels, and art works of all kinds which show that their knowledge of space, time and materials was more progressive, more fruitful and vastly more detailed than anything that emerged from the speculations of the philosophers. The speculations also suffered from internal difficulties. Thus the theoretical approach was not only useless, but incompatible with its own high standards of rigour.

From now on I shall call the knowledge desired by the early philosophers theoretical knowledge and traditions embodying theoretical knowledge theoretical traditions. The traditions to be superseded I shall call empirical or historical traditions. The members of theoretical traditions identify knowledge with universality, regard theories as the true bearers of information and try to reason in a standardised or 'logical' way. They want to bring knowledge under the rule of universal laws. Theories, according to them, identify what is permanent in the flux of history and thereby make it unhistorical. They introduce genuine, i.e. non-historical knowledge. The members of historical traditions emphazise what is particular (this includes particular regularities such as Kepler's laws). They rely on lists, stories, and asides, reason by example, analogy, and free association and use 'logical' rules when it suits their purpose. They also emphazise the plurality and, via the plurality, the history-dependence of logical standards.

The relation between the two traditions can be summarised in the following way:

(A) Historical traditions and theoretical traditions are both traditions in their own right, with their own laws, objects, research procedures and associated beliefs. Rationalism did not introduce order and wisdom where before there was chaos and ignorance; it introduced a special kind of order, established by special procedures and different from the order and the procedures of historical traditions.

24. *De generatione et corruptione* 325a 18ff, my emphasis.

(B) The theoretical approach ran into difficulties, both within itself and in the attempt to transform the historical traditions implicit in the crafts.[25] Most of these difficulties have survived until today; they have not been resolved. In religion we still have the clash between theologians who rely on an abstract notion of divinity and people desiring a more personal relation to god. In medicine we still have the clash between body theoreticians who judge disease from a single 'objective' point of view and clinicians who assert that a knowledge of disease presupposes personal interactions with the patient and the patient's culture.[26] Problems increase when we move from medicine via psychology to sociology, anthropology, history, philosophy.[27] Mathematics, which for Plato was *the* paradigm of theoretical knowledge, seems to return to the practical and 'subjective' philosophy of pretheoretical mathematicians: an increasing number of mathematicians and many computer scientists regard mathematics as a human activity, i.e. as a historical tradition. Theory is used, but in a free and experimental way. And the clash between the sciences and the humanities is nothing but a modern version of Plato's 'ancient battle'. All these antagonisms confirm thesis (A). They show that there was knowledge before there were theories, that this knowledge developed and improved, and that it has great staying power. They show that 'rationalism', the philosophy inherent in the theoretical approach, did not fully succeed in reducing the abundance of forms of knowledge that existed when it entered the scene and that a full reduction may do more damage than good. The modern debates about relativism

25. The idea that mathematics and astronomy owe their substance to philosophers — for example to Plato's proposal to develop astronomy from models, not from empirical observations — was criticized by O. Neugebauer, *The Exact Sciences in Antiquity*, New York 1962, p. 152. Cf. also his account of early Greek astronomy in Part Two of *A History of Ancient Mathematical Astronomy*, Berlin, Heidelberg & New York 1975.

26. According to Galen, *De sanitate tuenda*, I, 5, health is a condition 'in which we neither suffer pain nor are hindered in the functions of daily life'. But the functions of daily life are experienced differently by different people and they change from one culture to the next. Cf. O. Temkin, *The Double Face Of Janus*, pp. 441ff.

27. Anthropology has returned to the abundance philosophers and early scientists tried to overcome. Cf. the survey in G.E. Marcus and Michael M.J. Fischer, *Anthropology as Cultural Critique*, University of Chicago Press 1986. The essays in *After Philosophy*, B. Baynes, I. Bohman, Th.McCarthy, eds., Cambridge, Mass., 1987 seem to show that philosophy as a special subject with methods and a subject matter of its own is on the way out, but still fighting for survival.

and scepticism which define entire schools show, moreover, that theoretical traditions have not even succeeded in finding an adequate formulation of their own basic claims. The battle cry 'we need a new theory!' which is heard wherever a researcher or a whole discipline does not know what to do is, therefore, at most a party line supported by questionable arguments; it is not a necessary condition of knowledge.

(C) This, however, is not yet the end of the matter. For the difficulties and debates I have just mentioned shrink into insignificance when compared with the steady expansion of Western civilization into all areas of the world. I briefly described this phenomenon in the introduction. I also mentioned the enormous role played by science-based technologies. Thus theoretical traditions, while apparently defeated in words, and in peripheral domains of little power (philosophy, sociology, anthropology), seem to have won the battle in fact, and where it counts. Let us see what this victory amounts to!

5 On the Interpretation of Theories

Parmenides describes two procedures or 'ways of inquiry', as he calls them. The first procedure, 'far from the footsteps of humans', leads to what is 'appropriate and necessary'. The second procedure, based on 'habit, born of much experience' (*ethos polypeiron*: B7,3), i.e. on traditional attempts at gaining knowledge, contains the 'opinions of mortals.' According to Parmenides the first procedure, and it alone, establishes a truth capable of superseding all traditions.[28] Many scientists still believe that science can do the same.

The belief confounds the properties of ideas with their subject matter. According to Parmenides, statements such as '[Being] is' (which may be regarded as the first and most radical statement of a conservation principle) or 'it is homogeneous' (B8,22) describe the inherent structure of an entity that remains unaffected by human opinions. This is their subject matter. Similarly scientific statements are supposed to describe facts and laws that exist and

28. This becomes clear from his identification of Thought and Being (Diels-Kranz, fragment B3). The identification was implied in archaic Greece; Parmenides used it explicitly to combat the commonsense of his time. Scientists who emphasize facts and oppose speculation also believe that they can establish a tradition-independent truth.

govern events no matter what anybody thinks about them. However, the statements themselves certainly are not independent of human thought and action. They are human products. They were formulated with great care to select only the 'objective' ingredients of our environment but they still reflect the peculiarities of the individuals, groups, and societies from which they arose. Even the most abstract theories, though ahistorical in *intention* and *formulation*, are historical in *use*: science and its philosophical predecessors are parts of special historical traditions, not entities that transcend all history.

Thus the ancient trend towards uniformity which I described in the previous section, though supported by philosophers, was not initiated by them and did not lead away from history. As I explained in section 6 of chapter 1, it was part of an all-embracing historical development. Philosophers interpreted the development as the gradual emergence of a reality that had so far been concealed by ignorance and superficiality. The reality had always been there, they said, but it had not been recognized for what it was. They even believed that they had discovered it all by themselves, simply by using the powers of their amazing minds. For them the abundance of commonsense and of earlier traditions was not proof of an equally abundant reality, but of the protean nature of error. Parmenides represents an extreme case: reality has only one property, the property that it is, *estin* (B8,2).

In developing this theory Parmenides not only followed the trend, but was aided by the discovery (which may have been his own) that statements composed of simple concepts could be used to build new kinds of stories, soon to be called proofs, whose ending 'followed from' their inner structure and needed no external support.[29] The discovery seemed to show that true knowledge could indeed be used to judge traditions in a

29. Parmenides' poem contains a chain of arguments resulting in the denial of growth and decay (B8,6–21), parts and subdivisions (22–25), and movement (26–33). Each argument starts with the assertion to be denied, derives 'it is not' from it, and rejects the assertion on the basis of this consequence. (This seems to be the first explicit use of *reductio ad absurdum*. According to A. Szabo, *Anfänge der Griechischen Mathematik*, Budapest 1969, mathematicians learned *reductio* from Parmenides and his followers and accelerated the development of mathematics by its use. Traces of *reductio* can be found much earlier: see G.E.R. Lloyd, *Magic, Reason and Experience*, Cambridge 1979, pp. 59ff, esp. footnote 40.) Parmenides's arguments take it for granted that 'it is different' is the same as 'it is not' so that the only 'real' difference is between Being and Not Being (B8,16). This was suggested partly by the failures of the Ionians to identify a basic substance, partly by a strong 'existential' component of *einai*.

tradition-independent way. This is the error I mentioned before: the fact that simple ideas can be connected in simple ways does not change their nature; for example, it does not remove them from the domain of human activity. The power of the new ideas, therefore, did not reside in the ideas, in their connections and in the truths that emerged from them; it resided in the habits of those who, impressed by the increasing social and conceptual uniformity, preferred neat constructions to analogies; who, like Parmenides, were not overly interested in crude empirical matters; and who, unaware of the social roots of their disinterest, called such things unreal (Parmenides B2,6).

As opposed to Parmenides, his followers and some naive realists in science, the more sophisticated admirers of science admit and even emphazise that scientific theories are human creations and that science is one tradition among many. But they add that it is the only tradition that has succeeded in under-standing and changing the world. Theories, they say, played an important role in this twin achievement. They revealed an objec-tive order behind the confusing abundance of impressions and views, they provided points of attack for intended changes and they considerably reduced the amount of facts to be remem-bered. 'As science advances,' writes P.B. Medawar (*The Art of the Soluble*, London, 1967, p. 114),

> . . . particular facts are comprehended within, and therefore in a sense annihilated by, general statements of steadily increasing ex-planatory power and compass – whereupon the facts need no longer be known explicitly, i.e. spelled out and kept in mind. In all sciences we are being progressively relieved of the burden of singular instances, the tyranny of the particular.

Thus if we pay attention to some facts – the general features selected by science – we can disregard others.

This simple and rather popular account clashes both with scientific practice and with humanitarian principles. Social laws neither do nor should 'annihilate' the idiosyncracies 'of the parti-cular', i.e. of individual human beings. They do not annihilate them because each individual has features inaccessible to even the most comprehensive collection of laws – how else would people recognize each other as being different?[30] And they should not 'annihilate' them as that would violate the ideal of individual freedom cherished in many Western countries. The ideal is not universal and there exist societies where people try 'to stylise all aspects of personal expression to the point where anything idiosyncratic, anything characteristic of the individual merely because he is who he is physically, psychologically, or biographically, is muted in favour of his assigned place in the continuing and, so it is thought, never-changing pageant.'[31] But this only means that any 'annihilating' that occurs is the result of local customs and not of universal laws. Or to express it differently: social theory, though striving to supersede history, has only succeeded in becoming an uncomprehended part of it.

Similar considerations apply to the natural sciences. We may agree that a prediction of the path of Jupiter only needs its mass, velocity, location and the masses, velocities and locations of other relevant bodies. But this does not mean that the planet Jupiter has been absorbed into, or 'annihilated' by, celestial mechanics and has either ceased to exist as a separate entity or no longer contains excess information over and above the assertions of that theory. Jupiter also has non-mechanical properties, some of which are connected by non-mechanical laws: 'laws of Nature' can at most be regarded as abstractions in the sense in which Aristotle regarded mathematics as an abstraction (chapter 8); but abstractions are incapable of 'annihilating' anything.

Laws of nature are more than abstractions, the defenders of the non-historical nature of theory say; features such as mass, distance, velocity (in the case of classical celestial mechanics) are connected in ways that are independent of the interests of those who selected them. This separates them from other features (such as colours, or smells) which either lead to no connections at all, or to connections of a weak and casual kind. It shows that we are dealing with real properties and not merely with accidental objects of our curiosity. A researcher interested in reality accepts such properties, disregards (or 'annihilates') the remaining features of a situation and uses theories to guide him in the process of annihilation.

The argument takes it for granted that what is lawful or connected by laws belongs to a different level of existence than the

---

30. One of O. Sack's patients (he calls him Dr. P) was very good at identifying, and examining in his imagination, abstract patterns and schematic features, but he could not identify individual human faces. 'By a sort of comic and awful analogy,' writes Sacks (*op. cit.*, p. 20), 'our cognitive neurology and psychology [and, I would add, our sociology and politics] resembles nothing so much as poor Dr. P!'

31. Clifford Geertz, *Local Knowledge*, New York 1983, p. 62, describing the social constraints at work in Balinese society.

particular and idiosyncratic: it is part of a world that exists and develops independently of the thoughts, wishes, impressions of the researcher and is therefore 'real'. Now this assumption is not a result of research but of a metaphysics that separates Nature and Humanity, making the first stern, lawful and inaccessible and the second wilful, fickle and affected by the slightest disturbance. The metaphysics ceased to be popular long ago – but its epistemological shadow is still with us in the form of various versions of (scientific) realism. The shadow can be criticised by pointing out that connecting reality with lawfulness means defining it in a rather arbitrary manner. Moody gods, shy birds, people who are easily bored would be unreal, while mass hallucinations and systematic errors would become real.

We must also consider that the various points of view used to confer reality on some features and deny it to others do not form a coherent whole but conflict both with each other and with the evidence that allegedly supports them (the facts whose reality they are supposed to establish). The usual answer to this problem is that coherence can be achieved 'by approximation'. The answer is correct in some cases (the relation between classical mechanics and general relativity, for example), incomplete in others (the relation between quantum theory and chemistry – here the quantum theory is not merely curtailed, as in standard approximation procedures, but supplemented with new principles of a characteristically chemical nature) and meaningless in still further cases (botany or morphology can be related to molecular biology only by rejecting some of their basic features and declaring them to be 'unreal' or 'unscientific'): the 'reality' science allegedly defines and uses to 'annihilate' the more disorderly ingredients of our world is constantly being redefined to make it fit the fashion of the day.[32] Considering that the universe is now generally believed to have a history, that laws are said to arise as part of it and that we can only find evidence for those laws which, according to current belief, are necessary for life and consciousness,[33] we must suspect that even fundamental laws characterise particular stages of the world to a certain degree of

---

32. For details see Hans Primas, *Chemistry, Quantum Theory and Reductionism*, New York 1982, and Nancy Cartwright, *How the Laws of Physics Lie*, Oxford 1983.

33. This is one version of the so-called 'anthropic principle'. For a detailed discussion with literature cf. J.D. Barrow and Frank J. Tipler, *The Anthropic Cosmological Principle*, Oxford 1986.

precision but are not strictly true. 'The world is given to us only once', wrote Ernst Mach,[34] which means that statements implying ahistorical regularities are idealizations, or 'instruments', not descriptions of reality.

The attempt to make the success of science a measure of the reality of its ingredients fails also for other reasons. As I pointed out in Chapter 1 section 9, success and failure are culture-dependent notions: the 'green revolution' was a success from the standpoint of Western marketing practices, but a dismal failure for cultures interested in self sufficiency.

The argument remains in force when we disregard applications and deal with the theoretical validity of ideas instead. It is true that the validity of Maxwell's equations is independent of what people think about electrification. But it is not independent of the culture that contains them. It needed a very special mental attitude inserted into a very special social structure combined with sometimes quite idiosyncratic historical sequences to divine, formulate, check and establish the laws scientists are using today. This is now accepted by most sociologists, historians and philosophers of science. Now where is the argument to convince us that what arose in this highly idiosyncratic and culture dependent way, *exists* and *is valid* independently of it? What guarantees that we can separate the result from the way without losing the result? It needs only a minor modification of our technologies, ways of thinking and mathematics and we can no longer reason as we do within the status quo. And yet the objects of this status quo, the facts and laws we now regard as valid, are supposed to be 'in the world' independently of our thoughts and actions.

To support their assertion the 'realists' use the distinction, made a little earlier, between the subject matter of a statement, or a theory, or an approach, and the approach itself. Scientific statements, they say, are the results of a historical process. But this process happened to identify features of the world that are independent of it. As I argued in chapter 1, section 9, this line of reasoning can also be applied to the Greek gods. There is no way out: we either call quarks and gods equally real, but tied to different circumstances, or we cease to talk about real things altogether.

34. *Die Mechanik in ihrer Entwicklung*, Leipzig 1933, p. 222.

Note that such an interpretation does not deny the effectiveness of science as a provider of technologies and basic myths; it only denies that scientific objects *and they alone* are 'real'. Nor is it asserted that we can do without the sciences. The interpretation implies that we cannot. Having entered and elaborated an environment in which scientific laws come to the fore we, that is, scientists as well as the common citizens of Western Civilization are now subjected to their rule. But social conditions change and science changes with them. Nineteenth-century science denied cultural plurality, twentieth-century science, chastened by philosophical and practical failures (the failures of 'development' included) and by the invention of theories with decidedly 'subjective' ingredients, is no longer opposed to it. The same scientists, philosophers, politicians who want to increase the power of science by their very efforts transform science and the 'real' world with it. 'This world,' I said in chapter 1, 'is not a static entity populated by thinking ants who, crawling all over it, gradually discover its features without affecting them in any way. It is a dynamic and multifaceted entity which affects and reflects the activity of its explorers. It was once a world full of gods; it then became a drab material world and it will, hopefully, change further, into a more peaceful world where matter and life, thought and feelings, innovation and tradition collaborate for the benefit of all.'

These observations can be summarised in the following fourth point, to be added to the three points made at the end of the last section:

(D) Theoretical traditions are opposed to historical traditions in intention, not in fact. Trying to create a knowledge that differs from 'mere' historical or empirical knowledge, they succeeded in finding *formulations* (theories, formulae) which *sound* objective, universal and logically rigorous but which are *used* and, in use, *interpreted* in a manner conflicting with all these properties. What we have is a new historical tradition which, carried along by a sizeable false consciousness, seems to transcend human perception and opinion and human life. In this it shows great similarity with religious systems which have also transformed commonsense worlds and brought them closer to an otherworldly reality. Being supported by strong historical forces, the 'mechanization of our world picture' (Dijksterhuis) cannot be simply argued away. It needs strong counterforces to effect a change.

The counterforces exist. They are partly being mobilised by the aggressiveness of Western civilization, partly created within that civilization itself. Let us hope that they can overcome the dangers and drawbacks that accompany its sizeable achievements.

4
Creativity

1 Art and Science as Imitation

In Plato's *Timaeus* a being called God, or demiurg, or father builds the world by trying to make a disorderly and shapeless material conform to a precise and detailed plan. Like an engineer this God 'persuades' the material to yield 'the most excellent and perfect copy' of the plan. His achievement is the greater the closer the resemblance between plan and copy.

The Homeric poems contain set speeches (such as the speech given by a general to his troops before battle) and descriptions of typical states of affairs. The speeches and descriptions are repeated word for word when similar circumstances arise. The Homeric poets were not interested in producing new and 'original' expressions for the same old things. They wanted the best stereotype for a given situation and, having found it, they repeated it whenever the situation arose.

The idea that the arts repeat, or copy, or mimic reality in a different medium, using stereotypes of that medium as their building blocks, was the core of the ancient theory of *mimesis*. Plato's diatribe against the arts in book X of his *Republic* accepted the theory but criticized artists for imitating the wrong entities (physical objects or events instead of the principles to which both conform), for making deception (such as perspective) part of their imitative techniques and for arousing the emotions. He challenged the 'lovers of poetry to plead her cause and show that she is not only delightful, but beneficial to orderly government and human life.' Aristotle accepted the challenge; his answer, contained in his magnificent *Poetics*, remained within the framework of mimesis. Tragedy, he said, does indeed imitate; however, it does not imitate concrete historical events, as does history, it imitates underlying structures and is therefore 'more philosophical than history' (*Poetics* chapter ix). Tragedy is theory, history mere narrative. Numerous anecdotes about painted grapes that attract sparrows, painted horses that real ones neigh at, a painted curtain that fools even the artist's eye, and the countless epigrams about the deceptive lifelikeness of Myron's sculpted cow, show that in antiquity the imitative view was not just a philosophical specialty but part of commonsense.[1]

The view returned in the Renaissance. According to Leonardo 'that painting is most praiseworthy that has the most similarity to the thing reproduced, and I say this to refute the painters who want to improve upon the things of nature.'[2] Leon Battista Alberti, referring to the recently discovered rules of perspective, defined a picture as a 'cross section of the pyramid' formed by the rays that extend from the eye to the object presented.[3] Alberti's definition turns painting into the production of exact copies of cross sections of optical pyramids: 'I say the function of the painter is this: to describe with lines and to tint with colours on whatever panel or wall is given him similar observed planes of any body so that at a certain distance and in a certain position from the centre they appear in relief, seem to have mass and to be lifelike' (Spencer, *Alberti on Painting*, p. 89). Much later, in the nineteenth century, some artists tried to make Nature herself produce the copies, 'substitut[ing] her own inimitable pencil, [light,] for the imperfect tedious and almost hopeless attempt of copying a subject so intricate'. This led to the invention of photography.[4]

This brief survey already shows that imitation is not unambiguous but involves a series of choices. One is the choice of the material in which the copies are to be produced. The imitator must take the properties of the material into account. These properties may be due to laws of nature, they may be results of custom (the standard phrases, the grammar, the words of the

1. See Erwin Panofsky's essay *Idea*, New York 1986.
2. *Trattato della pittura*, ed. Ludwig, 1881, Nr. 411.
3. *Della Pittura*, quoted from J. R. Spencer, *Leon Battista Alberti on Painting*, New Haven and London 1966, pp. 52, 49.
4. Henry Fox Talbot, *Some Account of the Art of Photographic Drawing*, London 1893, quoted after Beaumont Newhall, *Photography, Essays & Images*, New York 1980, p. 27.

language used in written reports; metre, musical modes, standard gestures in tragedy), they may be inventions and traditions based on them (the second actor added to the drama by Sophocles and the third actor added by Euripides). Having chosen the material, the imitator must choose the aspects he wants to imitate. According to Aristotle both the historian and the writer of tragedy imitate – but they imitate different things. Writers of tragedy may further choose between lifelike actions without character or equally lifelike actions, but dominated by character.[5] Even in painting Aristotle distinguished between imitators like Zeuxis whose work was deceptive to the extreme but 'characterless' (1450a28) and others, like Polygnot, who imitated, but did not neglect character. Imitation is a complex process that involves theoretical and practical knowledge (of materials and traditions), can be modified by inventions and always involves a series of choices on the part of the imitator.

There exist philosophical views that make science a paradigm case of imitation. Imitation was part of Aristotle's theory of perception which, if undisturbed, imprints natural forms on the sense organs. It underlies the idea, widespread in antiquity and still popular today, that the task of science is to 'save the phenomena', i.e. to present them as correctly as possible using the available stereotypes (deferents, epicycles, equants and excentres in the case of Ptolemaic astronomy; differential equations in the case of classical physics). Imitation played a role in Bacon, who likened the mind to a bent and dirty mirror[6] whose surface had to be cleaned and made plane before it could produce true images of nature, and it survives in the popular image of the unprejudiced scientist who eschews speculation and concentrates on telling it like it is. However there exist other views which see the task of the sciences and the arts in a very different way.

According to one of these views which was introduced by Parmenides it is the task of (scientific) knowledge to describe reality. This sounds like imitation. However, Parmenides adds that reality is hidden behind deceptive phenomena and that divine support is needed to bring it to the fore. We have here a

---

5. According to Aristotle 'the moderns' (and that included Euripides) proceeded in the first way – 1450a25 – while 'nearly all the early men' (Aeschylus and Sophocles included) had greater success with character than with action – 1450a37.

6. *Novum Organum* Aphorism 47; cf. also aphorisms 115 and 69.

second account of the way in which the arts and the sciences are supposed to work.

2   Art and Science as Creative Enterprises

In antiquity the idea that the arts and the sciences imitate, that they are domains of rational judgement and that they can be taught was confronted by the view that 'poets do not create from knowledge but on the basis of a certain natural talent and guided by divine inspiration, just like seers and the singers of oracles.'[7] Speaking of poetry, Plato mentions 'being possessed by the muses, a madness that wakens a tender and untouched soul if it gets hold of it and makes it happy and educates by praising old stories in songs and in every other mode of poetry those who come after us. Whoever knocks on the door of poetry without the madness of the muses trusting that technique alone will make him a whole poet does not reach his aim; he and his poetry of reason disappear before the poetry of the madman' (*Phaedrus* 245a). In his Seventh Letter (341cf.), Plato explains how 'after a long continued intercourse between teacher and pupil, in joint pursuit of the subject, suddenly, like light flashing forth when fire is kindled, [knowledge of the ideas] is born in the soul and straightaway nourishes itself.'

In these passages the process of understanding or of building a work of art is said to contain an element that goes beyond skill, technical knowledge and talent. A new force takes hold of the soul and directs it, towards knowledge in the one case, towards a work of art in the other. If it is assumed that the force does not reach the individual from the outside, like divine inspiration or creative madness, but originates in the individual him(her)self and from there transforms the world (of art, of knowledge, of technology), then we have the idea of creativity I want to criticize in the present essay.

To make my criticism as concrete as possible I shall concentrate on a specific argument in its favour. And to make it as clear as possible I shall use an argument that tries to show the role of individual creativity in the sciences. If this clear and detailed argument fails then the rhetoric emerging from more foggy areas will altogether lose its force.

7. Plato, *Apology of Socrates*, 21d.

3   Einstein's Argument for Creativity Examined

In his essays 'On the Method of Theoretical Physics', 'Physics and Reality' and 'The Fundaments of Theoretical Physics' (republished in *Ideas and Opinions*, New York 1954 – page references will be to this book), Einstein explains why scientific theories and concepts are 'fictions' or 'free creations of the human mind' and why 'only intuition, resting on sympathetic understanding of experience, can reach them'. According to Einstein,

> . . . the first step in the setting of a 'real external world' is the formation of the concept of bodily objects and of bodily objects of various kinds. Out of the multitude of our sense experiences we take, mentally and arbitrarily, certain repeatedly occurring complexes of sense impressions (partly in conjunction with sense impressions which are interpreted as signs for sense experiences of others), and we correlate to them a concept – the concept of the bodily object. Considered logically this concept is not identical with the totality of sense impressions referred to; but it is a free creation of the human (or animal) mind. On the other hand, this concept owes its meaning and its justification exclusively to the totality of the sense impressions which we associate with it.
>
> The second step is to be found in the fact that, in our thinking (which determines our expectations), we attribute to this concept of the bodily object a significance, which is to a high degree independent of the sense impressions which originally give rise to it. This is what we mean when we attribute to the bodily object 'a real existence'. The justification of such a setting rests exclusively on the fact that, by means of such concepts and mental relations between them, we are able to orient ourselves in the labyrinth of sense impressions. These notions and relations, although free mental creations, appear to us stronger and more unalterable than the individual sense experience itself, the character of which as anything other than the result of an illusion or hallucination is never completely guaranteed. On the other hand, these concepts and relations, and indeed the postulation of real objects and, generally speaking, of the existence of 'the real world', have justification only in so far as they are connected with sense impressions between which they form a mental connection (p. 291).

Next, says Einstein, we introduce theories. Theories are speculative to a much higher degree. They not only are 'not directly connected with complexes of sense experiences' (p. 294), they are not even uniquely determined by observations – two different theories with different basic concepts (such as classical mechanics and the general theory of relativity) may fit the same empirical laws and the same observations (p. 273) – and they may clash with facts that were known at the time of their invention. The principles and concepts of theories are therefore entirely 'fictitious' (p. 273). Yet they are supposed to describe a hidden but objective real world. It needs a strong faith, a deeply religious attitude, to believe in such a connection and tremendous creative efforts are required to establish it.

This account of the growth of physical knowledge suffers from major difficulties. The starting point of the process which according to Einstein leads to reality is utterly unreal. There exists no stage in history, or in the growth of an individual, that corresponds to the 'first step'; there exists no stage when, surrounded by a 'labyrinth of sense impressions', we 'mentally and arbitrarily' select special bundles of experience, 'freely create' concepts and correlate them with the bundles. Even small children do not perceive bare colours and sounds, they perceive meaningful structures such as smiles or friendly voices. The perceptual world of the grown-up contains things and processes from tables and chairs to operatic performances, rainbows and stars. Most of these entities present themselves as being objective and independent of our wishes – we must push them, squeeze them, cut them to effect a physical change; a mere change of attitude or even of physical position does not suffice. Sense impressions such as pure colours and pure sounds (as opposed to coloured things and, say, human voices) play a negligible role in our perceptual world; they appear under special conditions that have to be carefully monitored (use of the reduction screen, for example); they are late theoretical constructs, not the beginnings of knowledge. Besides, it would not help us if they were: a person put into a 'labyrinth of sense impressions' could not possibly start constructing physical objects; he would be completely disoriented and unable to think the simplest thought. He would not be 'creative', he would be simply paralysed.

Let us now assume that the impossible has happened and that, having been given sense-data, we have succeeded in constructing, 'mentally and arbitrarily', a world of real objects. Are the further stages of Einstein's story correct? They are not!

It is true that a world of real objects is not *logically* equivalent to a selection of sense-data, however carefully arranged; however, it does not follow that this world has been constituted by a personal act of creation: when I walk, then the step I take now is not logically implied by the step I have just taken – but it would be silly to say that walking is the result of creative acts on the part of the walker. Or to take an example from inanimate nature: a later position of a falling stone does not follow logically from an earlier position; yet no defender of individual creativity would be inclined to say that the stone is falling creatively. Piaget has described in great and fascinating detail how the perceptions of children develop through stages without any conscious creative efforts on their part, simply obeying a 'law of evolution'.[8] Some features of our behaviour may even be genetically determined (for example, the recognition of patterns and the subsequent attachment to them, so beautifully described by Konrad Lorenz in his study of young geese). Result: the existence of a logical gap taken by itself does not yet show that it needs an individual creative act to bridge the gap.

Moving from commonsense to science, we encounter entirely new concepts: scientific theories are rarely expressed in everyday terms. They are formulated mathematically, different theories often using terms from different mathematical disciplines; the formalisms are connected with concepts and intuitions unfamiliar to commonsense; and the predictions, basic terms and theories are not uniquely determined by the known facts (this point was made a few paragraphs ago). The argument I am examining now concludes that the invention of the new languages and of the theories associated with them could not have occurred without considerable creative effort. Is *this* conclusion acceptable? I don't think it is; let me explain my reasons again.

My first reason is that the development of concepts need not be a result of the conscious actions of those using them. For example, the abstract notions of Being, Divinity, part and whole that were introduced by Xenophanes and Parmenides and elaborated by Zeno had been prepared by an unplanned gradual erosion of more concrete ideas. The erosion began in the *Iliad* and became noticeable by the sixth and fifth centuries B.C. The philosophers built on the erosion, they did not initiate it. The erosion affected behavioural concepts such as the concept of looking, social concepts such as the concept of honour and 'epistemological' concepts such as the concept of knowledge. Originally all these concepts included details of attitude, facial expression, mood, situation and other concrete circumstances. For example, there existed a concept of looking that contained

8. J. Piaget, *The Construction of Reality in the Child*, New York 1954, p. 352.

the fear felt by the person looking as an inseparable ingredient, and a concept of knowledge that incorporated the behaviour accompanying and propelling the acquisition of knowledge.[9]

The variety of the ideas then in existence, their complexity and their realism suggests even today that it may well be impossible to reduce our ways of being in the world to a few simple, context-(observer-) independent and therefore 'objective' notions. However, the number and the complexity of key concepts decreased, details disappeared, 'words became impoverished in content, they became one sided and empty formulae.'[10] Philosophers who prefer simple, clear and easily definable notions to complex, unclear and undefinable ones thrived on the deterioration and used it to assert, *after the event*, that there was essentially only one concept of knowledge, one concept of divinity, one concept of being. Thus a complex and detailed world view – the world view implicit in the Homeric epics – was replaced by a different, simpler and more abstract world view – the world view of the Presocratics (atomists included) and, then, of Plato – *without much conscious participation on the part of those who profited from the development*. (Later on, Aristotle restored important features of earlier thought and thus achieved an admirable synthesis of commonsense and abstract philosophy.) Of course, there were some minor 'discoveries' embedded here and there in the whole process – but they were of little importance and would have been useless without support from the major non-creative changes. Ernst Mach, one of the most imaginative philosophers of science, describes a similar situation in the history of numbers in the following way (*Erkenntnis und Irrtum*, Leipzig 1917, p. 327):

> One often calls the numbers 'free creations of the human mind'. The admiration for the human spirit that is expressed in these words is quite natural when we view the finished and imposing edifice of arithmetic. However our understanding of these creations is better served by tracing their *instinctive beginnings* and considering the circumstances which led to the need for these creations. Perhaps one will then realise that the first structures [Bildungen] which arose here were unconsciously and biologically *forced* upon humans by material

9. Details are found in Bruno Snell, *Ausdrücke für den Begriff des Wissens in der vorplatonischen Philosophie*, Berlin 1924, and in the same author's *Die Entdeckung des Geistes*, Göttingen 1975.
10. Kurt von Fritz, *Philosophie und Sprachlicher Ausdruck bei Demokrit, Platon und Aristoteles*, Darmstadt 1966, p. 11.

circumstances and that their value could be recognized only after they had proved useful.

My second reason for doubting that abstract and unusual concepts are exclusively the results of individual creative acts is that even the conscious and intentional formulation of novel general principles can be explained without resorting to creativity. An example will show what I mean. In chapter 1, section 2 of his *Mechanik* (9th edition, Leipzig 1933), Mach presents and analyses Stevin's argument for the conditions of static equilibrium on an inclined plane (equal weights on inclined planes of equal heights act in inverse proportion to the lengths of the planes – this I shall call proposition E). To derive E, Stevin imagines a chain suspended on a wedge containing the two planes. The chain, he argues, will either move or be at rest. If it moves, then it must be in perpetual motion (every position of the chain is equivalent to every other position). But a perpetual motion of the chain around the wedge, says Stevin, is absurd (proposition P). The chain is therefore at rest and in equilibrium. And as the lower parts of the chain, being symmetrical, can be cut off without disturbing the equilibrium, we obtain E. So far Stevin.

According to Mach, a proposition such as E can be found either by experiments and derivations therefrom, or with the help of 'principles' such as P. Experiments, he says, are 'distorted by alien circumstances (friction)', they 'always differ' from the 'precise static proportions', they 'appear doubtful' and the way from them to general laws is 'limping', 'unclear' and 'patchy' (op.cit., p. 72). Induction leads to sorry results. Arguments from principles, on the other hand, 'have *greater* value', and we 'accept them without contradiction'. The authority they possess derives from an 'instinct', for example from Stevin's conviction that the chain cannot possibly be in perpetual motion. This is the driving force of science for 'only the strongest instinct combined with the strongest conceptual power can make a person a great scientist.'

It is interesting to compare Mach's analysis of the use of principles in the sciences with Einstein's. Einstein read Mach with interest and was influenced by him in many ways. Following Mach, he did not start his paper on the special theory of relativity in the then usual way, by describing experimental results, but started with principles such as the principle of relativity and the principle of the constancy of the velocity of light. Throughout his life he mocked scientists who concentrated on 'measuring accuracies' and were 'deaf to the strongest arguments'.[11] Like Mach he thought certain facts too obvious to be in need of experimental support. For example, it appears that the Michelson-Morley experiment played no direct role in his construction of the special theory of relativity. 'I guess I just took it for granted that it was true', he replied to Shankland who had asked him precisely that question.[12] 'For it does not matter', Ernst Mach wrote in the case of Stevin, 'if one really carries out the experiment, provided the success is beyond question.' Stevin just carries out a thought experiment. 'And this', Mach says, 'is not a mistake – if it were a mistake, then all of us would share it.'

Mach's use of 'instinct' seems to bring us close to Einstein's 'free creations of the human mind' – but the difference is enormous. For while Einstein offers no analysis of the process of creation, preferring to connect it with his religious attitude, Mach immediately adds the qualification: 'This, however, forces us in no way to turn the instinctive elements of science into a new mysticism. Instead of practicing mysticism let us rather ask the following question: how does instinctive knowledge originate and what is contained in it?' And he replies that the instinct that makes a researcher formulate general principles without a detailed examination of relevant empirical evidence is the result of a long process of adaptation to which all of us, scientists as well as non-scientists, are subjected. Many expectations were disappointed during this process, behaviour changed accordingly and the human mind now contains the results of the changes. And as the number of disappointed expectations we experience in our daily lives and with it the number of confirmations for certain impossibilities (such as the impossibility that a chain may move with perpetual motion) is immeasurably greater than the number of the consciously planned experiments our scientists can perform, it is entirely reasonable to correct and perhaps even to suspend the results of such experiments with the help of instinctively found principles. Of course, Stevin's argument can start only after the two elements – the problem of the inclined plane and the instinctive knowledge concerning perpetual motion – have been brought together: Stevin must 'see' that the one can be solved by the other. But descriptions of scientific discovery[13] tell us that this 'bringing together' occurs almost by itself and is disturbed rather than helped by conscious interven-

11. *The Born-Einstein Letters*, New York 1971, p. 192.
12. *Am. Journ. Phys.*, Vol. 31 (1963), p. 55.

tion. Thus Mach has provided the elements of an explanatory sketch where Einstein (and Planck, and others) simply speak of 'free creations of the human mind'. The phenomena on which Einstein bases his view, taken by themselves, are therefore not yet proof of individual acts of creativity. We must take the analysis a few steps further.

4 The View of Human Beings that Underlies the Idea of Individual Creativity

I now come to my third and last comment on creativity. Speaking of creativity makes sense only if we view human beings in a certain way: they *start* causal chains, they are not just carried along by them. This, of course, is what most 'educated' people in the West assume today. They not only make the assumption, they also regard it as obvious. What else is a human being supposed to be? A human being has responsibilities, makes decisions, considers problems, tries to solve them and acts on the world in accordance with the solutions obtained. From our very childhood we are trained to connect events with our actions, to assume responsibility for them and to blame others for things we do not like. The assumption is the basis of politics, education, science and personal relations. However, it is not the only possible assumption and a life that rests on it is not the only form of life that ever existed. Human beings had (and in cultures different from ours still have) very different ideas about themselves, their lives, their role in the universe, they acted in accordance with these ideas and they achieved results we still admire and try to imitate.

As an example, take again the Homeric epics. A Homeric hero may find himself faced by various alternatives. Thus Achilles says, in *Iliad* book ix, lines 410ff (Lattimore tr.):

For my mother Thetis, the goddess of the silver feet tells me I carry two sorts of destiny towards the day of my death. Either if I stay here and fight beside the city of the Trojans my return home is gone, but my glory shall be everlasting; or if I return home to the beloved land of my fathers, the excellence of my glory is gone, but there will be a long life.

13. See the examples in J. Hadamard, *The Psychology of Invention in the Mathematical Field*, Princeton 1949.

Bruno Snell pointed out that passages such as these cannot be interpreted as saying that Achilles will *choose* the one path or the other; we must rather say that *he eventually finds himself on one of the two paths* and, having been given its description in advance, he now knows what he can expect: '. . . in Homer we never find a personal decision, a conscious choice made by an acting human being – a human being who is faced with various possibilities never thinks: it now depends on me, it depends on what I decide to do.'[14] And it could not be otherwise. Human beings, in Homer, simply do not have the unity needed for conscious choices and creative acts. Humans, as they appear in late geometric art, in Homer, and in archaic popular thought, are systems of loosely connected parts; they function as transit stations for equally loosely connected events such as dreams, thoughts, emotions, divine interventions. There is no spiritual centre, no 'soul' that might initiate or 'create' special causal chains, and even the body does not possess the coherence and the marvellous articulation given it in later Greek sculpture. But this lack of integration *of the individual* is more than compensated by the way in which the individual is embedded *into its surroundings*. While the modern conception separates the human being from the world in a manner that turns interaction into an unsolvable problem (such as the mind-body problem), a Homeric warrior or poet is no stranger in the world but shares many elements with it. He may not 'act' or 'create' in the sense of the defenders of individual responsibility, free will, and creativity – but he does not need such miracles to partake in the changes that surround him.[15]

With this I come to the main point of my argument. Today personal creativity is regarded as a special gift whose growth must be encouraged and whose absence shows serious shortcomings. Such an attitude makes sense only if human beings are self-contained entities, separated from the rest of nature, with ideas and a will of their own. But this view has led to tremendous problems. There are theoretical problems (the mind-body problem and, on a more technical level, the problem of induction; the problem of the reality of the external world; the problem of measurement in quantum mechanics, and so on), practical problems (how can the actions of humans who viewed themselves as the masters of Nature and Society, and whose

14. B. Snell, *Gesammelte Schriften*, Göttingen 1966, p. 18.
15. For details, cf. ch. 17 of my book *Against Method*, London 1975.

achievements now threaten to destroy both, be reintegrated with the rest of the world?), and ethical problems (have human beings the right to shape Nature and cultures different from their own according to their latest intellectual fashions?).

These problems are closely connected with the transition, already described, from complex and concrete to simple and abstract concepts. For while the earlier concepts took dependencies for granted and expressed them in various ways, the concepts of the 'philosophers', as the first theoretical scientists called themselves, and their seventeenth-century refinements were 'objective', i.e. detached from those who produced them and from the situations in which they were produced and therefore in principle incapable of doing justice to the rich pattern of interactions that is the world. It needs a miracle to bridge the abyss between subject and object, Man and Nature, experience and reality that is the result of these conceptual 'revolutions' – and creativity leading to wonderful castles of (philosophical and/or scientific) thought is supposed to be that miracle. Thus the allegedly most rational view of the world yet in existence can function only when combined with the most irrational events there are, viz. miracles.

5 Return to Wholeness

But there is no need for miracles. As I tried to show in my analysis of Einstein's argument, Einstein uses creativity for producing results (objects external to the observer; concepts more abstract than the commonsense concepts of a certain period) that are either transitory stages of a natural development (in the individual, in groups) or minute adaptations occurring in such stages. Neglecting the development and those features of a human being (of a group) that make it possible, Einstein starts from an abstract entity, the thinking subject, in fictitious surroundings, the 'labyrinth of [his] sensations'. Naturally he needs an equally abstract and fictitious process, creativity, to re-establish contact with real human beings and the results of their work. The gap that needs the miracle occurs in his model, it does not occur in the real world as described by researchers of a less abstract bent of mind (old fashioned biologists, non-behavioural psychologists) and by commonsense. Replace the model by this world and the spectre of individual creativity will disappear like a bad dream. Unfortunately, this is not yet the end of the matter.

The reason is that fictitious theories while out of touch with nature need not be out of touch with our beliefs and thus with culture. On the contrary, they often provide motives for strange and destructive actions. Unrealistic policies do not just collapse, they affect the world, they lead to war and other social and natural disasters. Once enthroned they cannot be easily dislocated by argument. Put into a hostile environment, the most beautiful argument sounds like sophistry – this is true of science, this is even more true of politics and of the commonsense that supports it in democratic countries. We do need arguments – but we also need an attitude, a religion, a philosophy or whatever you want to call such an agency, with corresponding sciences and political institutions, that views humans as inseparable parts of nature and society, not as their independent architects. We do not need new creative acts to find such a philosophy and the social structures it demands. The philosophy (religion) and the social structures already exist, at least in our history books, for they arose, long ago, when ideas and actions were still the results of a natural growth rather than of constructive efforts directed against the tendencies of such a growth. There are the Homeric epics, there is Taoism, there are the many 'primitive' cultures which put us to shame by their cheerful respect for the wonders of creation.

We cannot reject their views by claiming that they clash with 'science' or with 'the modern situation'. There is no monolithic entity, 'science', that can be said to clash with things, and 'the modern situation' is a catastrophe that offends our most basic desires for peace and happiness. Scientists themselves have started criticizing the separatist view of human beings, the view, that is, that there exists an 'objective' world and a 'subjective realm' and that it is imperative to keep them apart. Thus Ernst Mach pointed out, more than a century ago, that the separation cannot be justified by research, that the simplest sensation is a far reaching abstraction and that any act of perceiving is inextricably tied to physiological processes. Konrad Lorenz has argued for a science that makes 'subjective' factors parts of research, while one of the most advanced scientific disciplines, elementary particle physics, has forced us to admit that it is impossible to draw a sharp boundary between Nature and the agencies (mind included) used to examine her. Considering the social aspects, we need only remember the attitude of fifteenth-century Renaissance artists: they worked in teams, they were paid craftsmen, they accepted the guidance of their lay employers. Teamwork already plays an important role in the sciences; it was and still is exemplary in institutions such as the Bell Telephone Laboratories, leading to inventions (such as the transistor) which may well help us in our quest for a better world. All that is needed to restore the efficiency, the modesty and, above all, the humanity of the practitioners of a craft is the admission that scientists are citizens *even inside the domain of their expertise* and should therefore be prepared to accept the guidance and supervision of their fellow citizens. The conceited view that some human beings, having the divine gift of creativity, can rebuild Creation to fit their fantasies has not only led to tremendous social, ecological and personal problems, it also has very doubtful credentials, scientifically speaking. We should reexamine it, making full use of the less belligerent forms of life it displaced.

5
Progress in Philosophy, the Sciences and the Arts

1 Two Kinds of Progress

In a famous passage of book xxii of his *City of God*, St. Augustine gives a colourful description of human misery (I quote from the Modern Library Edition, New York 1950, pp. 847f):

> Who can describe, who can conceive the number and severity of the punishments which afflict the human race . . .? . . . we suffer robbery, captivity, chains, imprisonment, exile, torture, mutilation, loss of sight, the violation of chastity to satisfy the lust of the oppressor, and many other dreadful evils. What numberless casualties threaten our bodies from without – extremes of heat and cold, storms, floods, inundations, lightening, thunder, hail, earthquakes, houses falling; or from the stumbling or shying, or vice of horses; from countless poisons in fruit, water, air, animals; from the painful or even deadly bite of wild animals; from the madness which a rabid dog communicates, so that even the animal which of all others is most gentle and friendly to its own master, becomes an object of intenser fear than a lion or dragon, and the man whom it has by chance infected with this pestilential contagion becomes so rabid, that his parents, wife, children, dread him more than any wild beast! What disasters are suffered by those who travel by land and sea! What man can go out of his own house without being exposed on all hands to unforeseen accidents? Returning home sound in limb, he slips on his doorstep, breaks his leg and never recovers. What can be safer than a man sitting in his chair? Yet Eli the priest fell from his, and broke his neck . . . Is innocence a sufficient protection against the various assaults of demons? That no man might think so, even baptised infants, who are certainly unsurpassed in innocence, are sometimes so tormented, that God, who permits it, teaches us hereby to bewail

> the calamities of this life, and to desire the felicity of the life to come . . .

and so on for many more lines. This, says St. Augustine, is the 'just punishment' for the 'first sin'.

However God, being benevolent, also gave humans two precious gifts – the gift of procreation and the gift of invention. Invention has led to advances in all domains (*op.cit.*, p. 852):

> Has not the genius of man invented and applied countless astonishing arts, partly the result of necessity, partly the result of exuberant invention, so that the vigour of mind which is so active in the discovery not merely of superfluous but even of dangerous and destructive things, betokens an inexhaustible wealth in the nature which can invent, learn or employ such arts? What wonderful – one might say stupefying – advances has human industry made in the art of weaving and building, of agriculture and navigation! with what endless variety are designs in pottery, painting, and sculpture produced, and with what skill executed! What wonderful spectacles are exhibited in the theatres which those who have not seen them cannot believe. How skilful the contrivances for catching, killing and taming wild beasts! And for the injuries of man also, how many kinds of poisons, weapons, engines of destruction, have been invented while for the preservation and restoration of health the appliances and remedies are infinite! To provoke appetite and please the palate what a variety of seasonings have been concocted! To express and gain entrance for thought what a multitude and variety of signs there are, among which speaking and writing hold the first place! What ornaments has eloquence at command to delight the mind! What wealth of song is there to captivate the ear! How many musical instruments and strains of harmony have been devised! What skill has been attained in measures and numbers! With what sagacity have the movements and connections of the stars been discovered! Who could tell the thought that has been spent upon nature, even though, despairing of recounting it in detail, he endeavoured only to give a general view of it? In fine – even the defence of errors and misapprehensions which has illustrated the genius of heretics and philosophers, cannot be sufficiently declared.

In an equally famous passage of his *Lives of the Artists* Vasari notes that

> it is inherent in the very nature of the arts to progress step by step, from modest beginnings, and finally to reach the summit of perfection (Penguin Classics, p.85)

and he describes the recent achievements of this progress in the following way (op. cit., p. 83):

> The old Byzantine style was completely abandoned – the first step being taken by Cimabue and followed by Giotto – and a new style took its place; I like to call this Giotto's own style. In this style of painting the unbroken outline was rejected, as well as the staring eyes, feet on tiptoe, sharp hands, absence of shadows and other Byzantine absurdities. These gave way to graceful heads and delicate coloring. Giotto especially posed his figures more attractively, started to show some animation in the heads, and by depicting his draperies in folds made them more realistic; his innovations to some extent included the art of foreshortening. As well as this, he was the first to express the emotions, so that in his pictures one can discern expressions of fear, hate, anger or love. He evolved a delicate style from one that had been rough and harsh.

The two passages contain two important but different ideas of progress.

St. Augustine describes how humans enriched the arts and the sciences by developing new skills, styles, means of stimulating the mind and the senses, and even errors, and how the number of these skills, styles and so forth is constantly being increased. This *quantitative* or *additive idea* of progress is implied wherever an art or a science is praised for its inventions, discoveries, breakthroughs—for inventions, discoveries, breakthroughs are thought to be well defined and distinct events whose accumulation advances knowledge. The quantitative idea arose in antiquity; it is very popular today.

Vasari, on the other hand, gives a *qualitative account*. Progress, as he understands it, does not just increase numbers, it also changes the properties of things (skills, ideas, art works and so on).

The qualitative idea, too, has played an important role in history. It underlies the ancient story of different ages with different qualities of life (Hesiod, *Works and Days*, 109ff), it propelled the sciences, it influenced the arts and it is now being used with effect by advertising agencies ('Alpo dogfood – new and improved!').

The role of qualitative ideas of progress in the sciences is often concealed by a concern for quantitative details. But scientific debates about precision and numbers (of facts, or predictions) always involve qualitative assumptions which may be retained despite major empirical difficulties. The dispute between atomism and form theories, which continued right into the twentieth century, is an outstanding example. Atomism frequently conflicted with facts and reasonable (highly confirmed) theory. It survived because the idea that physical processes can be subdivided into elementary processes consisting in the motion of particles of matter was, in the minds of those holding it, stronger than the facts and the measurements that seemed to endanger it. The Copernican view, according to Copernicus, did not lead to more and better predictions than its rivals [1] but to a more harmonious account of the planetary system. Newton, who opposed intellectualised notions of god (such as Descartes's and Spinoza's) and emphasized personal relations between God and His creation, postulated continuing and active interactions between God and the material universe, while Leibniz turned God into a master architect who, having designed and created the most perfect world, let it develop in accordance with immutable laws. Observation and experiment played a small role in this debate which cut across the boundaries between physics, theology and religion. [2] And Einstein conceded the empirical success of the quantum theory while criticizing the conceptions on which it was based and the world view connected with them. Occasionally he went even further, ridiculing the widespread interest in a 'verification of little effects' and emphasizing the inherent reasonableness of his own point of view.[3] In all these cases (and, it seems, in all major scientific exchanges), qualitative assumptions play a decisive though often unnoticed role.

1. All the astronomical views which were available when Copernicus started writing were 'consistent with the data', as he says himself: *Commentariolus*, quoted from E. Rosen, ed., *Three Copernican Treatises*, New York 1959, p. 57.

2. For the general nature of the debate, cf. R.S. Westfall, *Science and Religion in Seventeenth Century England*, Ann Arbor Paperbacks 1973. Newton's religious views are described in Frank Manuel, *The Religion of Isaac Newton*, Oxford 1974. Leibniz's objections are contained in *The Leibnitz-Clarke Correspondence*, H. G. Alexander (ed.), Manchester 1956.

3. Cf. his reaction to Freundlich's measurements in his letter to Max Born, *The Born-Einstein Letters*, New York 1971, p. 192: 'Freundlich [whose observations seemed to conflict with the general theory of relativity] does not move me in the slightest. Even if the reflection of light, the perihelion, the lineshift [the three tests of general relativity known at the time] were unknown, the gravitation equations would still be convincing because they avoid the inertial system – the phantom which affects everything but is not itself affected. It is really strange but is not itself affected. It is really strange that human beings are normally deaf to the strongest arguments while they are always inclined to overestimate measuring accuracies.'

2 Their Different Properties

At first sight the two ideas seem to have different properties and
to work in different ways.

In its simplest form the quantitative idea seems to be an
'absolute' or objective idea: differences of opinion concerning the
number of objects of a certain kind cannot be explained by
differences of culture or purpose. We must concede that some or
all of the parties are mistaken – there exists only one correct
number for each assembly of things. It is this belief which apparently prompted Plato to prefer quantification to works of art:

> And have not measuring and numbering and weighing proved to be
> the most excellent aids against [errors and illusions] so that what is
> apparently larger or smaller or more or heavier does not prevail but
> calculation, measuring, and weighing? [*Rep*. 602d4ff].

The same belief lies behind the modern scientific drive for quantification. But numbers are obtained by counting. Counting
assumes that complex entities containing many parts (dogs,
operatic performances, novels, observations) are regarded as
units, and different people (different cultures) construct units in
different ways. How many constellations are there in the sky?
That depends on what figures are used to unite a collection of
stars into a single shape and how the elements of the shape are
connected with the whole. How many stars (globular clusters)
exist in our galaxy? That depends on the spectral range in which
you examine the matter: what looks like two different stars in
one range may turn into a single blob in another.

It makes no sense to say: 'But there must be a definite number
and we have to find it!', for the question is, 'A definite number of
what?' – and this 'What' changes with definition, manner of
viewing, theoretical assumptions and so on. Occasionally even
the number of well circumscribed entities depends on our
approach. How many people were in Capernaum after Christ
had arrived there? That depends on your view of Christ. If you
think that he was a human being, you will get one number. You
get a different number if you assume, as the docetists did, that he
was a mere appearance, a phantom body; and perhaps still
another number if you regard him as divine, without any human
admixture. The number of elementary particles in a given spacetime region depends on the nature of the interactions used to find
them. And so on. Only a persistent confusion of abstract

numbers (which seem clearly separable from each other) and numbers of objects (which depend on qualitative circumstances of the kind just described) could have made us believe that numerical judgements are more 'objective' than judgements of quality, structure, value.

Qualitative notions of progress, however, are *relative notions*: properties which are praised by some judges may be rejected by others. If all traditions disappear and only one tradition remains then the judgements of this one tradition will of course be the only judgements in existence – but they will still be relative judgements, just as 'bigger' remains a relation in a world with only one body in it. I shall now illustrate this feature of qualitative progress with examples from the arts.

3   Progress in the Arts

Vasari praises a natural posture, delicate colours and the bits and pieces of perspective he finds in Giotto. These are signs of progress – but only for a person who either consciously or by habit has adopted the view that a picture must repeat the *visible physical characteristics* of the object represented or, using more technical language already available in the fifteenth century, that it must reproduce the optical impressions of a well placed observer. For example, they are signs of progress for Leon Battista Alberti, who writes in his essay on painting:

> The function of the painter is this: to describe with lines and to tint with colours on whatever panel is given him similar observed planes of any body so that at a certain distance and in a certain position from the centre they appear in relief, seem to have mass and to be lifelike (*Leon Battista Alberti on Painting*, ed. John R. Spencer, New Haven and London 1956, p. 89).

In another passage, Alberti makes the geometrical aspect even more obvious: 'the picture is a cross section of the pyramid' reaching from the eye to the outlines of the painted object (op. cit., p. 52).

But the elements described by Alberti and praised by Vasari – perspective, natural postures, delicate colours, character, emotions – are *obstacles, not* improvements for an artist who wants a portrait or a statue to convey absolute power or spiritual eminence: what is permanent and independent of circumstances cannot be captured by what is transitory and relative to 'a certain distance and a certain position from the centre'. 'Natural postures' such as lying, standing, walking, looking around are particular and transitory situations, 'delicate colours' arise when pigments are immersed in a special light and viewed in a special atmosphere, emotions come and go and perspective, finally, does not give us an object, but how it looks to a perhaps quite insignificant individual: any bum can make an emperor as small as an ant by looking at him from a distance. Artists interested in power, permanence and objectivity have been aware of these shortcomings and have developed special methods which no longer conform to the principles of an extreme optical realism.

The history of Egyptian art is an excellent historical example; here an early and quite sophisticated naturalistic style was replaced by a stern formalism that remained unchanged for centuries (Plato, *Laws* 256d f, commented favourably on the stability achieved). But the naturalistic techniques were not forgotten. They were used for scenes from everyday life and they returned in all domains during the reign of Amenophis IV (see my *Against Method*. Chapter 17). This means that deviations from a 'faithful rendering of nature' occurred in the presence of detailed knowledge of the objects represented and side by side with more realistic accounts. We can here watch, almost as in a laboratory, how styles and methods of representation change with the purpose for which the art works are designed.

Another example which shows that some apparently 'primitive' procedures may be the results of purpose and not of ignorance or lack of talent is the early history of Christian art. Catacomb paintings seem to have been guided by two aims: *to decorate* and *to tell a story*. The building blocks of the stories were pictures – but they functioned more like the pictographs in early Chinese writing than like the snapshots of criminals in a police station. Individual expressions, perspective, delicate colouring are lacking not because we are in a 'primitive stage' but because it would be nonsensical to demand these features from a *sign* signifying 'face', or to request that a sign meaning 'house' be constructed in accordance with the rules of perspective. (Similarly, it would be nonsensical to demand individual expressions from the drawings of a human figure designed to explain the neural connections between the retina and the brain.) Even portraits of individuals (saints, bishops, emperors), such as statues or pictures (on panels, on coins) inscribed with proper names, often wore conventional expressions because their purpose was to

> convey the idea that the person portrayed is the real basileus, consul, dignitary, or bishop by showing in the portrait that he possesses all the essential characteristics: he has noble and grave features and a majestic mien, he makes the correct gesture, he holds in his hand the insignia or he wears the clothes appropriate to his social situation. We are tempted to say, when we see for instance an image of St. Theodore as a soldier, that it is an image of a Byzantine soldier which resembles other portrayals of Byzantine soldiers. But what we should say is that it is an image of St. Theodore defined inconographically as a Byzantine soldier (André Grabar, *Christian Iconography*, Princeton 1968, pp. 65f.).

The problems an optical realism poses for an attitude of this kind are stated very clearly in a Greek text known as the Apocryphal Acts of John and attributed by philologists to the second century A.D. and to Asia Minor. The text tells us how Lycomedes, a pupil of John, secretly invited a painter to John's house and asked him to paint a portrait of John. John discovered the portrait but, never having seen his own face, failed to recognize it and thought it to be an idol. Lycomedes brought a mirror and John, comparing mirror and painting, said:

> As the Lord Jesus Christ lives, this portrait is like me, yet, my child, not like me but only like my fleshly image. For if this painter who has here imitated my face wants to draw it in a portrait, he will be at a loss [needing more than] the colours, which you now see and the boards . . . and the position of my shape and old age and youth and all the things that are seen with the eye.
>
> But you, Lycomedes, should become a good painter for me. You have the colours which he gives you through me, he, who himself paints all of us, even Jesus, who knows the shapes and the appearances and the postures and types of our souls. But what you have done here is childish and imperfect: *you have drawn a dead likeness of the dead* (Grabar, ibid., pp. 66f.; my emphasis).

In other words: optical realism leaves out life and the soul.

Pseudo-John's diagnosis receives support from the psychology of perception which shows that the information conveyed by a naturalistic rendering of an object is burdened by redundancies which interfere with the attempt to grasp its structure.

Ryan and Schwarz ('Speed of Perception as a Function of Mode of Representation', *American Journal of Psychology* Vol. 69 (1956), pp.

60ff) compared four modes of representation:(a) photographs; (b) shaded drawings; (c) line drawings; and (d) cartoons of the same objects. The pictures were presented in brief exposures, and the subject had to specify the relative position of some part of the picture, e.g. the positions of the fingers in a hand. The exposures were increased from a point at which they were too brief, until they were sufficient for accurate judgements to be obtained, with this finding: Cartoons were correctly perceived in the shortest exposure; outline drawings required the longest exposures; and the other two were about equal, and fell between these extremes (quoted from Julian Hochberg, 'The Representation of Things and People', in E.H. Gombrich, Julian Hochberg and Max Black, *Art, Perception and Reality*, Baltimore and London 1972, pp. 74f).

We may infer that the details of a naturalistic rendering of a face similarly delay recognizing the 'character', or the 'soul', or the 'essence' of the individual portrayed, and that character or essence may present themselves in a way not only concealed but perhaps not even touched upon by optical realism. The last part of the inference is supported by further discoveries about the nature and development of perception. 'Gestalt psychology', writes Anton Ehrenzweig in his book *The Hidden Order of Art* (Berkeley and Los Angeles 1967, p. 13),

predicted that on opening their . . . eyes to the world [the] attention [of formerly blind people] would at once be attracted to forms displaying . . . basic patterns [such as spheres and circles, cubes and squares, pyramids and triangles]. What a unique opportunity for observing the gestalt principle at work in automatically organizing the visual field into a precise figure seen against an indistinct background! None of these predictions came true! Case histories collected by von Senden (M. von Senden, *Space and Light*, London 1960) show the incredible difficulties encountered by patients as they were suddenly faced with the complexities of the visual world. Many of them . . . faltered in their purpose and could not muster the effort needed for organizing the buzzing chaos of coloured blotches. Some of them felt profound relief when blindness overcame them once again and allowed them to sink back into their familiar world of touch.

They showed neither a great facility nor inclination for picking out basic geometric shapes. In order to distinguish, say, a triangle from a square, they had to 'count' the corners one by one, as they had done by touching them when they were still blind. They often failed miserably. They had certainly no immediate easy awareness of a simple self-evident gestalt as the gestalt theorists had predicted. Simplicity of pattern played only a small part in their learning. The psycho-analysts will not be surprised to hear that a libidinous interest in reality rather than abstract form was the greatest incentive and the most efficient guide. A girl who was an animal lover identified her beloved dog first of all. A recent case showed that the face of the physician was the first shapeless blob picked out from the general blur of the visual field.

Thus optical realism is at least twice removed from the reality of a person: it is overloaded with details and it relies on the projected shapes of an object instead of its emotional impact. Primitive artists, modern painters such as Picasso or Kokoschka, the author of the apocryphal acts of John and the many unknown artists of the early Christian period seemed to understand this situation far better than their realistic critics. (The latter, on the other hand, may have been influenced by the alienation that increasingly pervades our lives.)

That the elements described by Alberti and praised by Vasari – perspective, natural postures, delicate colours and so on – might be obstacles, not achievements, becomes very clear when we consider the function of *colours*. The bright colours criticized by Vasari – the ultramarine of the heavens, the golden splendour of the haloes, the shining reds and greens of the garments – played an important role in mediaeval art: being intentionally anti-naturalistic, they lent to the light of the picture an extramundane essence, as a modern historian of colouring describes the matter (Schöne, *Über das Licht in der Malerei*, Berlin 1954, p. 21). Abbot Suger of St. Denis described this 'anagogical' or 'upward leading' function of antinaturalistic luxury while contemplating the precious stones on the main altar of his church (*Liber de Administratione*, xxxiii, quoted from E. Panofsky's edition in his *Abbot Suger*, Princeton 1979, pp. 63f):

> When – out of my delight in the beauty of the house of God – the loveliness of the many coloured stones has called me away from external cares, and worthy meditation has induced me to reflect, transferring that which is material to that which is immaterial, on the diversity of the sacred virtues: then it seems to me that I see myself dwelling, as it were, in some strange region of the universe which neither exists entirely in the slime of the earth nor entirely in the purity of heaven; and that, by the grace of God, I can be transported from this inferior to that higher world in an anagogical manner.

I conclude that the history of the arts presents us with a variety of techniques and means of representation employed for a variety of reasons and adapted to a variety of purposes. The attempt to diagnose progress across all reasons and purposes would be as silly as an attempt to interpret the diagrams in Gray's anatomy and a crucifixion on a rural road as stages of a single and ascending line of development. Some painters at the time of Vasari, and Vasari himself in particular, did notice the difference of approach. Lacking historical perspective they did not realise the change of purpose connected with it; they assumed that the new aims they themselves pursued had always been the aims of painting; accordingly, they interpreted every step towards these aims as progress and every step away from them as decay – a simple mistake which lies at the basis of much of the discomfort felt about relativistic views. (An interesting optical manifestation of the mistake is the development of the halo: first it surrounds the saintly heads in golden splendour, then it slowly becomes elliptical, until it finally turns into a veritable ring of Saturn; there exists also a 'materialistic' rendering where the circular back of a chair signals the holiness of the person sitting in it.)

4 Philosophy

At first sight the situation in philosophy seems to differ considerably from the situation in the arts. Philosophy is the domain of thought and thought seems to be objective and independent of styles, impressions, feelings. Now, to start with, this is itself a philosophical theory. There are other views, such as that of Kierkegaard, who asserts that thought receives content by being connected with a thinker, is essentially subjective and is incapable of producing 'results' — that is, permanent and unchanging signposts for an evaluation of the evanescent opinions of humanity. 'While objective thought', writes Kierkegaard,

> translates everything into results and helps all mankind to cheat, by copying these off and reciting them by rote, subjective thought puts everything in process and omits the results; partly because this belongs to him who has the way and partly because as an existing individual he is constantly in process of coming to be, which holds true of every human being who has not permitted himself to be deceived into becoming objective, inhumanly identifying himself with speculative philosophy in the abstract (Kierkegaard's *Concluding Unscientific Postscript*, D. F. Swenson and Walter Lowrie, eds., Princeton 1941, p. 68 – the reference is to Hegel; but it might as well be to the science-ridden philosophies of today).

According to Kierkegaard we have a choice: we can start to think objectively, produce results but cease to exist as responsible human beings, or we can eschew results and remain 'constantly in the process of coming to be': different forms of life have different philosophies. Bohr, who studied Kierkegaard's ideas in detail (cf. Niels Bohr, *Collected Works*, Vol.1, Amsterdam 1972, pp. 500f.) was averse to freezing even precise and well confirmed facts and points of view:

> He would never try to outline any finished picture, but would patiently go through all the phases of the development of a problem, starting from some apparent paradox and gradually leading to its elucidation. In fact, he never regarded achieved results in any other light than as starting points for further exploration. In speculating about the prospects of some line of investigation, he would dismiss the usual considerations of simplicity, elegance and even consistency . . . (Rosenfeld in S. Rozenthal, ed., *Niels Bohr, His Life and Work as Seen by his Friends*, New York 1967, p. 117).

Bohr's writings, accordingly, are heavily seasoned with historical material and are best characterised as preliminary summaries, surveying the past, giving an account of the existing state of knowledge and making suggestions for future research.

Secondly, important changes of abstract thought are qualitative changes, and qualitative changes, whether of thought or of art works, are inherently relative. Indeed, we might interpret philosophy as an art, like painting, or music, or sculpture — the difference being that while sculpture works with stone or metal, painting with colours and light and music with sounds, philosophy works with thoughts, bends them, connects them, cuts them up and builds fantastic dream castles out of this airy material. My earlier sketches (in chapter 1, section 6 and chapter 3, section 4) of the development (in the West) from pre-philosophical thought to philosophy show that this interpretation is very plausible indeed: the transition from the Homeric world view to the Presocratics and especially to Parmenides' philosophy of the One, though aided by large scale social developments, leads to a situation in which we again have a choice: we can accept the new monotony and adapt our lives to it, or we can regard it as being 'next door to lunacy' (Aristotle, *De Gen. et Corr.*, 325a18f) and continue relying on commonsense. The theoretical sciences went the first way. History, natural history included, the arts and the humanities tried hard to choose the second. Again it does not make sense to arrange all philosophies on a single progressive line.

5   The Situation in the Sciences

The arts and philosophy have tried, and some artists and philosophers still are trying, to overcome relativism. They have not succeeded and they cannot succeed from the nature of the case: qualitative preferences have no inherent order. The theoretical sciences try to establish such an order by subjecting qualitative judgements to the laws of quantitative progress: ideas leading to a greater number of successful predictions are 'objectively' better ideas. [4] Assume the attempt succeeds. Then the sciences could be characterised as those arts which, using not colours, not metals, not sounds, not stones, but ideas, are not only *talking about* progress but *generate it*, and in a manner that must be acknowledged by all. I conclude my comments on progress with a critique of this idea, in four points.

To start with, the combination of quantity and quality that allegedly characterises the sciences is itself a qualitative idea and therefore not absolute. If I have a friend then I shall want to know lots of things about him but my curiosity will be limited by my respect for his privacy. Some cultures treat Nature in the same respectful-friendly way. Their whole existence is arranged accordingly, and it is not a bad life, neither materially, nor spiritually. Indeed one asks oneself if the changes resulting from more intrusive procedures are not at least in part responsible for the ecological problems and the pervasive feeling of alienation we are facing today. But this means that the transition from non-science to science (to express a highly complex development in terms of a simple alternative) is progressive only when judged from inside a particular form of life. (It should also be noted that the actual sciences, as practiced by scientists, have little to do with the monolithic monster 'science' that underlies the claim of progressiveness.)

Secondly, conditions guaranteeing an increase of predictions often lead to qualitative problems which raise serious questions about the reality of the increase. Modern science rejects qualities, but it relies on them in its observation statements: every

4. This is a rough characterization of an idea that has led to extended technical debates. The objections below do not depend on the details of these debates.

observation statement is burdened by the mind-body problem. This does not affect scientists who regard theories as mere calculating devices; it is a difficulty for scientific realists. Some thinkers, Berkeley, Hume and Mach among them, took the problem seriously. Most scientists are either unaware of it or push it aside as a minor philosophical puzzle. This means they limit the domain of natural knowledge and define importance and unimportance only with respect to, or relative to, what happens inside the limits. Aristotle was not content with such facile manouevres.

Thirdly, the transition from one theory to another occasionally (but not always) involves a change of all the facts, so that it is no longer possible to compare the facts of one theory with those of the other. The transition from classical mechanics to the special theory of relativity is an example. The special theory of relativity does not add new non-classical facts to the facts of classical physics and so increase its predictive power, it is incapable of expressing classical facts (though it can provide approximate relativistic models for some of them). We must start all over again, as it were. Entire disciplines (such as the classical theory of the kinematics and the dynamics of solid objects) disappear as the result of the transition (they remain as calculating devices). Professor Kuhn and I have used the term 'incommensurability' to characterise this situation. Moving from classical mechanics to relativity we do not count old facts and add new facts to them, we start counting all over again and therefore cannot talk of quantitative *progress*.

Fourthly and finally, the qualitative elements of the sciences or, what amounts to the same thing, the fundamental ideas of a certain branch of knowledge are never uniquely determined by the facts of that branch. And by this I don't just mean that, given any set of facts, there always exists a variety of theories that agrees with the set; I rather mean that even a refuted rival of a theory that is highly confirmed and 'scientifically sound' (whatever that means at the time of the judgement) may overtake the successful theory: research may remove evidence from it and transfer it to the disreputable rival while at the same time using the evidence that makes the rival disreputable to discredit the successful point of view. Thus the free fall of heavy bodies for a long time supported the idea that the earth was at rest. Experience teaches that motion needs a moving force and comes to a standstill when the force stops acting. Now if I release a stone from the top of a tower, and if the earth moves, then it no longer follows the moving earth, it stays behind and should describe an inclined trajectory. The stone falls straight down – hence the earth is at rest. Galileo replaced the Aristotelian law of motion – each motion needs a moving force, otherwise there is no motion – by his own highly speculative law and was now obliged to transfer the evidence in favour of Aristotle to his own view. This he did not succeed in doing – the theory of friction and of air resistance, and the whole subject of aerodynamics, did not exist at the time. But he started a process in whose course the arguments in favour of an unmoved earth were gradually defused and transferred to the Copernican point of view. Similar developments accompanied the reemergence of the wave theory of light during the nineteenth century and the survival of the atomic theory throughout the anti-atomist period of the nineteenth century. In fact, we can say that the battle between alternative qualitative points of view, being redefined whenever new ideas and instruments of combat (experimental procedures, mathematical techniques) enter the scene, never really comes to an end and that support for one side of the battle can never he shown to be 'objectively misguided'.

However, it can be withdrawn for other reasons such as lack of patience, lack of funds, a firm conviction that the chosen way is the right way or a preference for the line of least resistance.[5] Such a choice, though not arbitrary, has no 'objective', i.e. opinion-independent, support. Still, it may affect many people. Major aspects of Western societies are the results of 'subjective' choices of this kind. In a democracy subjective choices with major implications are in the hands of the citizens. Of course, the citizens are not always better informed than the scientists (though many citizens, having a wider view than specialists, are able to spot areas not covered by any expertise) – but in the cases we are now discussing they are no worse informed either. They may not know all the detailed evidence that supports a generally accepted scientific doctrine and they may be completely unaware of the

5. 'The great success of Cartesian method and the Cartesian view of nature is in part a result of a historical path of least resistance. Those problems that yield to the attack are pursued most vigorously, precisely because the method works there. Other problems and other phenomena are left behind, walled off from understanding by the commitment to Cartesianism. The hard problems are not tackled, if for no other reason than that brilliant scientific careers are not built on persistent failure': see R. Levings and R. Lewontin, *The Dialectical Biologist*, pp. 2f. ('Cartesianism' here means the same as reductionism.) This observation applies to many fields, the quantum theory among them.

sophisticated arguments based on that evidence, but they are asked to decide not about the character and the strength of this support but about the chances for a comeback of unpopular alternatives – and on this point the scientists are equally in the dark. They are right to conjecture that if they continue the battle in the familiar manner they are not likely to lose, but they cannot say what will happen when different weapons get the same financial support they themselves enjoyed in the past. The so-called authority of the sciences, however, i.e. the use of research results as barriers to future research, relies on decisions whose correctness can only be checked by what the decisions eliminate – a typical feature of totalitarian thought. Further material on this point can be found in Ch. 1, section 2, text to footnotes 16 and 17.

Comments on a Discussion of this Paper

An earlier version of this paper was read (by professor Günther Stent – I was not present at the Symposium) at the Nobel Symposium 58, held at Lidingoe, Sweden from August 15 to August 19 1983. The title of the Symposium was Progress in Science and Its Social Conditions; the Proceedings containing that version were published by Pergamon Press in 1986. A brief discussion and a final summary produced a variety of critical points. In what follows I comment on some of them.

On incommensurability, it was pointed out that 'concepts such as "motion", "velocity", "acceleration" [were] refined so that problems posed at previous stages of development find an answer at later stages'. This is true in some cases, untrue in others. As I show in chapter 8, the 'classical' notion of continuity propagated by Galileo and still accepted by Hermann Weyl was not a refinement of the Aristotelian notion, but a considerable simplification. Also, the refinements that did take place concerned only certain (not all! — cf. my remarks, above, about continuity) aspects of locomotion. Other types of motion simply dropped out of sight.

It was also pointed out that there exist many relations between, say, classical and relativistic notions. That is true – but the relations are of a purely formal kind. What is important for me is that in accepting the basic postulates of relativity we must admit that classical notions cease to be applicable (gods and molecules have certain formal features in common – they can both be counted – but this does not mean that gods can be reduced to, or subsumed under, the principles of a mechanical materialism). 'There exists a certain continuity' – yes, if one does not look too closely and especially if one rests content with formal relations. Scientists who are practical people don't look too closely. But the myth of progress was introduced by philosophers; *they* insist on precision, *they* must admit that the development of the sciences contains many discontinuities.

Qualitative notions are not *in themselves* relative, it was pointed out, and the example was given of a cube of ice which melts (qualitative process) when put into a sauna. Correct – but I spoke of qualitative notions *of progress*, and these, containing evaluations, are always relative.

There were some critical remarks about my suggestion, which I made in a few lines towards the end of the paper, that science should be subject to democratic control. 'Feyerabend says nothing about how these councils are to be composed and chosen', was one objection. True. This was an aside in a paper devoted to other matters. If I went into details I would say that it is not up to *me* to describe the structure and the function of the councils, but up to those who introduce and use them: democratic measures are *ad hoc*, they are introduced for a specific purpose and serve specific people, so their structure cannot and should not be determined by distant theoreticians (see sections 5 and 6 of chapter 12). The tired ghost of Lysenko was called up, too, to protect science from public control. But the Lysenko affair did not happen in a democracy, it happened in a totalitarian state where special groups (conservative scientists and politicians) and not the people as a whole decided about scientific matters. Giordano Bruno was not burned by democratic councils either, but by experts. Besides, Lysenko made some good points in opposition to the one-sided genetic predictions of his time.

True, there was violent resistance against impressionism, expressionism, cubism and so on. What was wrong with that? Resistance does no harm as long as it is not insitutionalised, and here again the institutionalised resistance was not that of democratic councils, but of older academic schools. Of course, a democratic supervision of science may eliminate things loved by some scientists, but note that in the present situation scientists can eliminate things loved by non-scientists. It simply is not possible to have every single wish fulfilled. In these circumstances it seems wise to consider the opinions of all the people affected by a particular programme, idea, point of view, and not only those of a small elite. And, of course, experts will not be excluded. They will have ample opportunity to present their suggestions and to explain why so much money is needed to have them realised. As regards the question of whether I myself would survive in such circumstances – well, we just will have to see!

A final comment on 'progress'. 'It seems to me quite obvious', says a critic, 'that we know more about the world than people did in the days of Parmenides and Aristotle.' Well, that sounds very nice and plausible – but who is the 'we' the critic is talking about? Is he talking about himself? Then the statement is quite obviously false – there is no doubt that Aristotle, on many subjects, knew more than he does. On certain subjects he knew even more than the most advanced scholars of today (for example, he knew more about Aeschylus than any modern classical scholar). Is the 'we' the 'educated layman'? Again the statement is false. Is the 'we' all modern scientists? Then there are many things which Aristotle knew but which modern scientists don't know and from the nature of their business cannot possibly know. The same is true when we replace Aristotle by Indians, or by Pygmies, or by any 'primitive' tribe that has succeeded in surviving plagues, colonization and development. There are lots of things unknown to 'us' Western intellectuals but known to other people. (The reverse is of course also true – there are lots of things unknown to others but known to us. The question is: what is the balance?)

It may be true that the sum total of the facts that now lie buried in scientific journals, textbooks, letters and hard discs by far exceeds the sum total of the knowledge that comes from other traditions. But what counts is not number but usefulness and accessibility. How much of this knowledge is useful, and to whom? In a letter published in the January 1987 issue of the *Notices of the American Mathematical Society*, James York writes that a study by Eugene Garfield of the one thousand most cited scientists 'turned up *zero* mathematicians' (report by Gina Kolata in *Science* 1987, p.159). Mathematicians don't cite each other very often either, and non-scientists almost never pay any attention to them. Bulk of material is one reason, specialist jargon another. Research papers are not read by all the authors who sign them and are read with little attention by others, as is shown by the fact that trivial mistakes can survive for years and are often only discovered by accident. Most of the 'knowledge' that is sitting around is as unknown as were quarks around the turn of the century. It is there, but it is not and cannot be consulted and 'we' certainly do not know it. So looking at the matter in somewhat greater detail and with greater specificity than is suggested by the vague 'we' of the critic, we find lots of problems, but no obvious answers. It is therefore necessary to go beyond empty slogans and to start *thinking*.

6

Trivializing Knowledge: Comments on Popper's Excursions into Philosophy

The three books I shall comment upon (*Realism and the Aim of Science*, to be abbreviated as R; *Quantum Theory and the Schism in Physics*, to be abbreviated as Q – these two books are part (vols. 1 and 3) of Popper's *Postscript* (to the *Logic of Scientific Discovery*); and *Auf der Suche nach einer besseren Welt*, Munich 1984, to be abbreviated as S) are collections of essays of varying length, written between the early fifties and the late eighties. Some of the essays had been published before and were reprinted with minor modifications; others are new. Once more Popper explains his philosophy and tries to overcome 'the general anti-rationalistic atmosphere which has become a major menace of our time' (Q 156). I shall concentrate on three topics — critical rationalism, falsification and realism, and quantum theory — and I shall conclude my review with a brief account of the recent history (over the last hundred years) of the philosophy of science and of the role of rationalism in general.

1. Critical Rationalism

Critical rationalism, the 'real linchpin of [Popper's] thought' (R xxxv), is a tradition he himself has traced back to the Presocratics and especially to Xenophanes. The tradition is rational, it 'wishes to understand the world and to learn by arguing with others' (R 6). It is pluralistic – the arguments play points of view off against each other rather than comparing them with a fixed source of knowledge. It favours democracy which, according to Popper, is a form of society 'that can be changed by words and, now and then, though rarely, even by rational argument' (S 130). And it regards scientific achievements as the most important events in the history of mankind (S 208, on Newton; Q 158, on Einstein). 'I opt for Western Civilization, science, democracy,' Popper wrote in an essay read on the occasion of the 25th anniversary of the Austrian treaty of independence (S 130).

For Popper the transition from 'closed' societies that rest on relatively stable institutions, customs and beliefs to 'open' societies which examine every aspect of the world is a step in the right direction. Enriching existence by a discussion of its ingredients, societies which take the step 'may succeed, by [their] own critical efforts in breaking down one or another of [their] prison walls' (R 16) – they may succeed in removing traditional boundaries of thought and action.

But the traditions that contain the boundaries give meaning to the lives they restrict. The movement towards an open society is, therefore, not without difficulties. There are gains, there are losses. Popper himself has painted a vivid picture of the losses that occurred in ancient Greece. In the final chapter of Volume 1 of his *Open Society*, he speaks of the 'burden of civilization', of the 'sense of drift' experienced by those entering it, and describes the uncertainty and the loss of meaning caused by the gradual movement towards an 'abstract society' that reduces personal contacts and increases the distance between man and nature. However, he shows little sympathy for those who, noticing the difficulties, have tried to alleviate them – such attempts, he claims, are symptoms of immaturity: the burden is the price we have to pay for becoming human. And he adds that people and societies unwilling to pay the price might have to be forced to abandon tribal habits, just as the ancient Greeks were forced, by 'some form of imperialism'.

These are not merely academic disputes or historical reflections. Changes analogous to the ones described by Popper are going on right before our eyes. I am of course talking about the steady expansion of Western civilization into all areas of the world; and I am especially talking about a more recent part of this expansion which has been called, somewhat euphemistically, 'developmental aid'. The expansion was, and still is, ruthlessly imperialistic – yet many of the countries behind it are now run in a more or less democratic manner. This means that for these countries at least the extent and the quality of the 'aid' are

in principle subjected to a democratic vote: *we ourselves* are
called upon to decide if and how we should interfere with the
lives of strangers. What our governments are offering are the
fruits of science and civilization and the means of increasing
them. According to Popper this is 'the best' (S 129) humanity has
produced. Should we let the recipients choose, and perhaps
familiar manner, by 'some form of imperialism'?
return, or should we, following Popper, regard rejection as a sign
of immaturity and impose our own mature will in the old and
familiar manner, by 'some form of imperialism'?

In an essay published in 1981, Popper wrote: 'Concerning any
value to be realised by a society, there exist other values that
collide with it' (S 129). Again all change, change in the direction
of a critical rationalism included, is said to be accompanied by
gains as well as by losses; however, gains and losses are now
described in a more 'objective' manner. They are said to correspond to 'values'. But if there are positive elements on both
sides, then statements such as 'for me our Western Cicilization is
the best' (S 129) are subjective opinions about their relative
weight, and dissenting views cannot be dismissed as signs of
immaturity.

We can go even further. 'Everywhere in the world,' writes
Popper (S 128f.), 'humans have created new and often vastly
differing cultural worlds: the worlds of myth, of poetry, of art, of
music; the worlds of means of production, of tools, of technology, of economy; the worlds of morals, of law, of protection
and aid for children, for sick people, for the weak and for others
in need of help.' According to Popper such worlds are the results
of numerous trials and errors extending, on occasion, over thou-
sands of years. They are forms of knowledge that have passed
many tests and are corroborated to a high degree (S 17ff.). The
same applies to 'theories [that] are incorporated in our language;
and not only in its vocabulary, but also in its grammatical structure' (R 15) — to all the tribal cosmologies, that is, which
Benjamin Lee Whorf explored with such consummate skill (R
17). Popper also asserts that it is 'reasonable' or 'rational' to rely,
'*for all practical purposes*', on what has been severely tested and
has survived the tests (R 62, original emphasis, the example
being Western science). It follows that it is 'reasonable, or
rational' not only to *accept* the views implicit in cultures different
from our own – this was the argument of the last paragraph – but
to *rely on* them, 'for all practical purposes' (i.e. for the treatment
of problems that arise in their midst), rather than on the

'dreams . . . indulge[d] in' (Q 177) by philosophers (and distant 'developers'), and on the untested 'bold hypotheses' they are producing with such abandon. The early Greek critics of Presocratic speculation said the same.

Thus Herodotus and Sophocles wrote about the anthropomorphic Greek gods as if Xenophanes (who criticized them from an ahistorical point of view) had never existed, and Herodotus also gave arguments for their power; the sophists who were the foremost political thinkers of their time continued to deal with morals in the loose, semi-empirical ways they knew from poetry, especially from Homer, while Plato, the theoretician par excellence, could never completely live without the older and more 'primitive' ideas and customs (cf. his frequent change from argument to myth and back to argument again). The reaction of the crafts, or *technai* as they were then called, is especially interesting. Herodotus, the 'father of history' (Cicero) and one of the first geographers, ridiculed Hekataeus who had tried to incorporate geographical information into Anaximander's 'bold' sketch of the world. He wrote (*Histories* iv, 36):

> I must laugh when I see how people nowadays draw maps and give explanations devoid of reason; they draw an ocean flowing around the earth and the earth itself they make round like a circle formed on a potter's wheel and Asia they make as big as Europe.

Nature, Herodotus said, is a little more complex than that. We have seen (Chapter 1, section 6) that the author of *Ancient Medicine* makes a similarly ironic comparison between the suggestions of the Presocratics and the medical practice of his time. An enormous amount of useful empirical knowledge was contained in, and defended against theoretical reductions by, the remaining arts and crafts (cf. again chapter 1, section 6). Thus it was not only 'positivism' that 'opposed speculation' (Q 172), but a vastly more substantial tradition which prepared the ground from which the dreams of the philosophers and the later dreams of a theoretical science could grow (I am now talking only about Western philosophers and especially about the ancient Greeks. Chinese speculation, it seems, always stayed close to the practice of the crafts). The tradition did not avoid unobservable entities (its world is not a 'world without riddles' – R 103); however, it respected familiar distinctions and well known facts and its 'intellectuals' opposed the presumption that such facts counted little, were mere 'subjective appearances', when compared to the 'deeper reality' allegedly revealed by the theoreticians. The tradition was, and its modern successors still remain, 'empirical' in precisely this sense and not in the sense of being tied to a special 'source'. The tradition is conservative, for it prefers institutions and information that have passed the test of time to the 'dreams . . . indulge[d] in' by arm-chair philosophers (cf. Q 177 – dreams are not rejected; but they are not made the centre of civilization either). It uses induction, i.e. it lets practice guide thought rather than opposing thought to practice. And it is rational: it uses arguments against its light-footed alternative. Some of the ancient arguments were quoted above; Professor von Hayek's plea for a free market and against government interventions and his related protest against the disruption of established social institutions continue them into the present. Most interesting is Mach's idea (*Mechanik*, Leipzig 1933, pp. 25ff) that far reaching speculations and 'principles' of a high degree of abstraction introduced by scientists with a strong 'instinct' are successful because they are firmly embedded in the empirical reality they try to explain: 'instinct' is the result of numerous (mostly unconscious) trials and errors; it is corroborated to a high degree ('it tells us what cannot occur', says Mach). Experimental results and empirical regularities have much smaller support; hence, they can be corrected 'from above', with the aid of principles (we can correct them while explaining them, says Popper: R 144).

The view (which plays an important role in Popper's writings) that all forms of knowledge are the result of trial and error can therefore accomodate (at least) two different traditions which I shall call theoretical traditions and historical traditions respectively. In the West theoretical traditions were closely connected with the rise of philosophy and theoretical sciences such as mathematics, astronomy and basic physics, while historical traditions contained the arts (in the ancient sense of *technai*) and other forms of practical knowledge. Popper does not criticize the latter; he criticizes 'positivism', a school philosophy of no importance. This is due to his tendency to reduce historical antagonisms to simpleminded alternatives. Popper also advises researchers to use 'bold hypotheses', that is, hypotheses which not only go beyond but also against accepted facts – he clearly prefers theoretical traditions. In his *Poverty of Historicism* he argues that they alone count. Let us examine this assumption!

Historical traditions (which contain the humanities, the arts both in their ancient and in their modern sense, as well as the so-called '*Geisteswissenschaften*') produce knowledge that is restricted, either explicitly or by use, to certain regions and depends on conditions specifying these regions; they produce *regional* or, with respect to the conditions, *relative knowledge* (of what is good or bad, true or false, beautiful or ugly, etc.). Herodotus' story of Darius (*Histories* iii, 38), which Popper quotes with approval as evidence for the critical attitude of early Greek thinkers (S 134), makes precisely this point: *customs* change from one society to the next, they are 'relative to' the societies that have them – but this does not make them meaningless, or reduce their force as Popper implies (S 216f). On the contrary, 'only a madman would mock them' (Herodotus, same passage, not quoted by Popper). Cambyses, who broke into temples, burned holy images, desecrated ancient graves, examined the corpses that had been buried in them and mocked customs unfamiliar to him, was not, according to Herodotus, an enlightened thinker, he 'was completely mad'. Protagoras (cf. the great speech of the character 'Protagoras' in Plato's dialogue of the same name, 322d4f and 325b6ff), Plato (*Theatatus* 172a, reporting Protagoras' doctrine that 'what is good and bad, just and unjust, pious and impious is what a state thinks it is and then declares to be law') and, much later, Montaigne and his followers in the Enlightenment said the same.

The early geographical, medical and ethnographic literature applied this point of view not only to customs but also to the world at large: different countries have different shapes and climates, plants and animals change from one region to the next, there are different races with different ideas about the world they inhabit and different ways of making these ideas plausible. The entire universe consists of regions or domains, each characterised by a special 'climate' and special laws. Poseidon's objection against the universalistic pretensions of Zeus (*Il*. 15, 184ff) made the point in the imagery of the epic, while Aristotle replaced the *physical regions* (which were the ancestors of the elements) by *theoretical subjects* (poetry, biology, mathematics, cosmology) with concepts, laws and conditions of their own. Combining the fact (already known to Homer: *Od*. 18, 136) that human insight is changed by (physical and social) circumstances with the realization (present in Herodotus) that even the strangest customs and beliefs are essential parts of the lives of those who have them, and aid them in various ways, we arrive at the view that *all opinions, though relative or regional, are worth considering*. Herodotus accordingly gives equal weight to the achievements of Greeks and Barbarians. At the start of his great historical work, he writes:

> This is what Herodotus of Halikarnassos has found out so that the things done by humans do not fade away and the great and astonishing works, be they now produced by Greeks or by Barbarians do not remain unreported.

Later Greek chauvinists (Plutarch, for example) had no sympathy for such a wide perspective.

Theoretical traditions, on the other hand, try to create information which no longer depends on, or is 'relative to', special conditions and which is therefore 'objective', to use a modern term. Regional information, in these traditions, is either disregarded, or it is pushed aside, or it is subsumed under comprehensive points of view and thus changed in its nature. By now many intellectuals regard theoretical or 'objective' knowledge as the only knowledge worth considering. Popper himself encourages the belief by his slander of relativism (S 216).

Now this conceit would have substance if scientists and philosophers looking for universal and objective knowledge and a universal and objective morality had succeeded in finding the former and persuaded, rather than forced, dissenting cultures to adopt the latter. This is not the case. As I argued in chapter 1, section 5, and again in chapter 2, the regionalism of phenomena was never overcome, neither by science nor by philosophy. What we have are modest successes in narrow domains and grandiloquent promises dressed up as results already achieved. It is true that Popper is opposed to reductive attitudes; he has asserted that 'realism . . . should be at least tentatively pluralistic' (*Objective Knowledge*, Oxford 1972, p. 294; cf. p. 252). According to him, there are

> . . . many sorts of real things . . . foodstuffs . . . or more resistant objects . . . like stones, and trees, and humans. But there are many sorts of reality that are quite different, such as our subjective decoding of our experiences of foodstuffs, stones, and trees, and human bodies . . . Examples of other sorts in this many-sorted universe are: a toothache, a word, a language, a highway code, a novel, a governmental decision; a valid or invalid proof; perhaps forces, fields of forces . . . structures . . . (p. 37).

But the entities just mentioned can be parts of the same real world only if the theories that constitute them (most 'modern' entities such as electrons, quarks, light signals, spacetime regions are 'theoretical entities' and so are the entities, already referred to, of tribal cosmologies) can be readily united – and that is not the case. The difficulty is not formal (although there are such difficulties as well) but connected with the fact ('incommensurability') that some of the views to be united, when used, deny conditions of use for the statements of others (see chapter 17 of my *Against Method*, London 1975, esp. pp. 269ff., as well as p. 15 of Vol. 1 of my *Philosophical Papers*, Cambridge 1981); that is, it is again connected with their 'relative' or 'regional' character. Quantum mechanics, the theory dominating the 'middle region' of the universe, *even contains* the idea of relative knowledge (complementarity). All this does not bother practicing scientists: they have no compunction about combining bits and pieces of different theories in a manner that would give a heart attack to purists. For them, science is not a theoretical tradition expressed in 'deductive system[s]' (Q 194), as Popper assumes, but historical tradition in the sense just specified – which brings me to my next topic.

2   Falsification and Realism

Critical rationalism leads to a subdivision of forms of life or cultures into those that tend to examine all aspects of their existence and others that leave certain aspects untouched. Within the former, Popper has developed a more specific theory about the difference between scientific and non-scientific conjectures and the nature of scientific change. The theory goes back to the *Logic of Scientific Discovery*. In R, p. xixff., Popper restates it and defends it against his critics. He asserts that the 'theory was not intended to be a historical theory, or a theory to be supported by historical or other facts' (R p. xxxi; cf.p. xxv), but he adds (italics in the original): 'yet I doubt that there exists any theory of science which can throw so much light on the history of science as the theory of refutations followed by revolutionary and yet conservative reconstruction.' I shall now examine this claim, starting with Popper's criterion of demarcation—falsifiability.

According to Popper, 'divisions of learning are fictitious and badly misleading' (R p. 159); there can, therefore, not be 'any sharp demarcation between science and metaphysics; and the significance of the demarcation, if any, should not be overrated' (p. 161). For example, 'even pseudo-sciences may well be meaningful' (p. 189, title line). But there are two reasons why it is not entirely futile to talk about demarcation, one theoretical, one practical.

The theoretical reason concerns problems of the 'logic of science' (p. 161)—i.e, of a domain of knowledge *about* rather than *of* the sciences. Here it may be admitted (though some alleged 'inductivists' will protest) that Popper's criterion of falsifiability is at least logically possible, while 'inductivist' criteria (in Popper's sense) are not: given a theory and a class of statements it is indeed 'a matter of pure logic' (p. xxi) to decide if the theory is falsifiable relative to the class *provided* both the theory and the statements are formulated in the language of a particular logical system (with implicit assumptions spelled out in detail) and have received a well defined interpretation. Scientific theories and experimental statements as used by scientists do not conform to the proviso. They are never completely formalised or fully interpreted, and the class of basic statements is never simply 'given'. Now we may treat theories as if they did conform to the proviso – in which case falsifiability is shown not of real scientific theories relative to real experimental reports but of caricatures with respect to other caricatures. On the other hand, we may use scientific theories as they are used by scientists, in which case the content of both theory and experiment is often *constituted by* the refutations performed and accepted by the scientific community rather than being the *basis on which* falsifiability can be decided and refutations carried out: one withdraws a theory because of certain difficulties and thereby decides what kind of theory one wants it to be. Popper is inclined towards the first alternative, the caricatures, which means that he 'may . . . legitimately be treated as a naive falsificationist' – to use Kuhn's words (cf. R, p. xxxiv).

'The problem of demarcation [between science and non-science]', Popper continues, not only has a theoretical side, 'it is also of considerable practical importance' (R p. 162); it provides means of redirecting research: given an influential view with many followers and many successes to its credit, it may be fruitful to invite the followers to look for falsifying instances (pp. 163ff.). But it may be equally fruitful to emphasize the support for a theory threatened by difficulties and not unambiguously 'scientific' (i.e. falsifiable). Most of the arguments for the atomic theory were of this kind, and it was their collective weight that kept the theory alive; the same is true of Newton's gravitational theory from Newton's own time up to Laplace (the problem of perturbation and especially the problem of the great inequality of Jupiter and Saturn). Emphasizing falsifiability is therefore only *one* helpful move among many in the game of science (this goes beyond the 'anarchism' of R pp. 5ff., which is restricted to the *invention* and the *truth* of theories).

Next comes falsification. As before, Popper stresses the 'uncertainty of every empirical falsification', adds that this uncertainty 'should not be taken too seriously . . . there are a number of important falsifications which are as "definitive" as general human fallibility admits' (R p. xxiii), and calls a 'legend' the assertion 'that falsification plays no role in the history of science. In fact,' he says (p. xxv), 'it plays a leading role.'

It is not easy to evaluate the last statement. 'Leading' may have a quantitative meaning (falsifications vastly outnumber other events), or a qualitative meaning (no important development without falsification) or both (most important developments are brought about by falsifications). I shall argue against Popper on the basis of the last interpretation (parallel arguments can be found for the other two cases). And my argument is that to establish the leading nature of falsification in this sense would require a knowledge of the *percentage* of revolutionary theoretical changes brought about by refutations among all revolutionary theoretical changes, as well as decisions on which changes are theoretical and revolutionary and which are not. No information exists on the first point and there is lots of leeway on the second: for some historians of astronomy Copernicus was a revolutionary while for others, like Derek de Solla Price, he was a conservative; for some scientists Einstein's special theory of relativity was, and still is, 'the relativity theory of Poincaré and Lorentz with some amplifications', as E. Whittaker wrote, while for others it was and still is a bold new point of view. Yet despite all these difficulties I think it is possible to throw doubt on Popper's claim.

To start with, there are many cases where major clashes between theory and fact are recognized and pushed aside as an irritating interference with the process of research which then produces important discoveries. An example is the fate of Kepler's and Descartes's rule that an object viewed through a lens is perceived at the point of intersection of the rays travelling from the lens towards the eye. The rule connected theoretical optics with vision and gave it an empirical basis. The rule implies that an object situated at the focus will be seen infinitely far away. 'But on the contrary,' wrote Barrow, Newton's teacher and predecessor at Cambridge, 'we are assured by experience that [a point close to the focus] appears variously distant, according to the different situations of the eye. . . . And it does almost never seem further off than it would be if it were beheld with the naked eye; but, on the contrary, it does sometimes appear much nearer . . . All which does seem repugnant to our principles.' 'But for me,' Barrow continues, 'neither this nor any other difficulty shall have so great an influence on me, as to make me renounce that which I know to be manifestly agreeable to reason.' And so the situation remained, until the nineteenth century. The only thinker troubled by the conflict (and encouraged by it to develop his own philosophy) was Berkeley – see his *Essay towards a New Theory of Vision*. The attitude is very common and has prevented the premature modification of useful points of view. (References and further examples are found in my book *Against Method*, London 1975, Chapter 5.)

Let us now examine Popper's own argument. He offers a list of decisive refutations (R, p. xxvi). But what we need is not enumerative induction, but an estimate of percentages (see above), and such an estimate is nowhere found in his work. The list itself tells an interesting story that has little to do with what Popper extracts from it.

Not all items on the list are instances of refutation. Thus Galileo (item 2) refuted special explanations which Aristotle gave for special kinds of motion — for example, he refuted the theory of antiperistasis; however he did not refute *but accepted* Aristotle's general theory (he accepted impetus). He dropped impetus when he introduced what is now known as Galilean relativity (and which he never formulated clearly and consistently). Aristotle's general theory was never *refuted*; it *disappeared* from astronomy and physics but continued to aid research in electricity, biology and, later, epidemiology. Toricelli (item 3) did not refute 'nature abhors a vacuum' – no experimental investigation could have done that (how do you show, by experiment, that there is nothing in the space you are looking at? The space at least contains light, as Leibniz remarked in his debate, via Clarke, with Newton). Guericke's *Experimenta Nova* very clearly show the difficulties of the matter. Guericke promises to 'silence empty talk by letting the facts speak as witnesses'; he discovers that no space can be completely voided of matter; he ascribes this to the 'effluvia' emitted by all objects; and he conjectures that these effluvia will stay close to the earth so that there must be a vacuum somewhere in interstellar space.

A nice argument (which, incidentally, assumes what is to be shown, namely that matter consists of atoms with nothing in between) – but is it a refutation? Newton saw the problem and he used planetary theory as an argument against a full space. This gives us a lower limit for density, but no vacuum, unless we interpret low density as occupation of but a small part of space, i.e. unless we again already assume a vacuum.

A second difficulty with Popper's list is that cases that seem to fit the pattern of refutation followed by reconstruction are often complex events with refutation as a minor, almost trivial, and certainly not a 'leading' ingredient. Atomism (item 1) is an excellent example. According to Popper, 'Leucippus takes the *existence of motion* as a partial refutation of Parmenides' theory that the world is full and motionless' (p. xxvi). This cannot possibly be the whole story! It suggests that Parmenides, being too engrossed in speculation, overlooked motion and that Leucippus found what Parmenides had overlooked and used it to refute him. But Parmenides of course knew that there was motion – in the second part of his poem he even gave an account of it – but he regarded it as unreal. He sharply distinguished between truth and reality on the one side and 'habit, based on manifold experience' on the other, and he banned motion from the former. He thereby anticipated a prominent feature of the sciences: they, too, restrict what is real to a special domain and dismiss 'subjective' events such as feelings, perceptions, and so on.

Now the decision about what is to be regarded as real is one of the most important decisions an individual or a group can make – it affects the private and the public lives of everybody. Hence, the wish to retain a certain form of these lives may favour some decisions over others. In Aristotle this social or 'political' element of 'epistemological decisions' about reality and appearance becomes very clear. Arguing in a manner reminiscent of the author of *Ancient Medicine*, he writes about the reality of a universal Good (*Nicomachean Ethics*, 1096 b32ff.):

> Assuming there exists a Good that is one and can be predicated of everything or that exists separately and in and for itself, it is clear that this Good can be neither realised nor acquired by humans. But it is such a good [i.e. that can be acquired by humans] that we are looking for –

that is, we are looking for things that play a role in our lives. And the question is: *should we adjust our lives to the inventions of specialists or should we adapt these inventions to the requirements of our lives*? Parmenides (and Xenophanes whose divine being, far from superseding anthropomorphism, is a monstrous and power-hungry super-intellectual) chose the first way. Leucippus (Q, p. 162) and Aristotle chose the latter. (So does Popper. For him 'realism is linked with . . . the reality of the human mind, of human creativity and of human suffering' (Q, p. xviii), and 'any argument against realism . . . ought to be silenced by the memory of the reality of the events of Hiroshima and Nagasaki': (Q, p. 2). Leucippus seems to have proceeded in a rather intuitive way, whereas Aristotle made the principles of choice explicit (cf. also book 1 of his *Physics* for a criticism of Parmenides). The choice having been made, the 'refutation' is an afterthought, not a 'leading' element of the transition from Parmenides to the atomists. This is true of many other cases on Popper's list.

A third difficulty is that it 'often takes a long time before a falsification is accepted' (R, p. xxiv) and that acceptance occurs *as a result* of theoretical changes and upheavals which according to Popper are *caused* by the falsification. Popper has an inkling of the situation, for he writes that falsifications 'usually [are] not accepted until the falsified theory is replaced by a proposal for a new and better theory' (ibid.). The photoelectric effect (item 12) is an excellent example (relevant material is found in Bruce Wheaton's thesis *The Photoelectric Effect and the Origin of the Quantum Theory of free Radiation*, Berkeley 1971).

According to Popper, 'Philipp Lenard's experiment's . . . conflicted with what was to be expected from Maxwell's theory' *(R, p. xxix)*. For whom? — ' . . . as Lenard himself insisted', writes Popper. Wrong! For Lenard, the experimental findings he had assembled by 1902 (saturation current independent of light intensity; noticeable influence of the 'type of light' – but no quantitative relation between frequency and the energy of the ejected electron) did not present the slightest difficulty. He regarded them as an indication of complex processes going on inside the metal surface and he welcomed the photoelectric effect as a tool for examining these processes: 'This result', he wrote (*Annalen der Physik* Vol. 4 (1902), p. 150), 'suggests that in the process of emission the light plays only a role of triggering motions which must exist permanently with full velocity within the body atoms.' (The 'triggering theory' was called the 'modern theory' until at least 1910). Einstein's paper of 1905 contains some interesting speculations, a precise prediction, but no refutation. Calculating the entropy of monochromatic radiation for low radiation density from the 'false' law of Wien, he found it analogous to the entropy of a gas consisting of bundles of energy. From that he deduced an equation about the photoelectric effect that went beyond what had so far been found by experiment. In 1914 Millikan interpreted the equation as entailing three assertions, viz. (1) there exists a linear relation between maximal energy (stopping potential) and frequency; (2) the value of the slope of this line is h/e for all metals; (3) the intercept of the line gives the threshold frequency of emission – and he confirmed all of them for a sodium sample. But neither he nor Planck nor even Bohr were prepared to regard Maxwell's equations as refuted. Bohr especially stuck to classical wave theory until the early thirties – and with good reasons. Millikan expresses the general attitude: 'Experiment has outrun theory, or, better, guided by erroneous theory, it has discovered relationships which seem to be of the greatest interest and importance, but the reasons for them are as yet not at all understood' (*The Electron*, Chicago 1917, p. 230). Einstein himself, at the first Solvay Conference of 1911 (*Proceedings*, Paris 1912, p. 443), described his ideas as follows: 'I insist on the provisional character of this concept which does not seem reconcilable with the experimentally verified consequences of the wave theory': the wave theory was not endangered by quanta (or by the photoelectric effect), quanta were endangered by the wave theory. The particle character of light (and the refuting nature of the photoelectric effect) were accepted only after the discussions of the interpretation of the quantum theory had come to a preliminary end, which means that the photoelectric effect became a refutation *after* the processes its falsifying character allegedly intitiated had run their course. The same is true of the Michelson experiment (item 9), of the 'refutation' of anti-atomistic views, of Thomson's electron. Almost all Popper's examples, when studied by historians of science relying on documents, not 'mainly on [their] memory' (p. xxvi), change from prominent refutations leading to major theoretical reconstructions into processes where refutations play a rather uninteresting secondary role. They do occur, but they are not the prime motor of scientific change. Once again Popper, who thinks they are, 'may . . . legitimately be treated as a naive falsificationist'.

Popper is a realist – 'Realism is the message of this book', he writes about Q (p. xviii). He takes his conception of reality from the (Western) sciences and from (Western) commonsense. Now scientific realism – the idea that there exists a world independently of us which we can explore in a critical way – contains a component similar to Parmenides' distinction between true knowledge and opinion based on habit or experience. Like this distinction or boundary, the distinction drawn by the realists in Western science can be moved by practical decisions (cf. the above discussion of Parmenides). The anti-realistic views Popper attacks are based on some rather academic versions of such decisions: they emphasize certainty and they put the boundary between sense-data and the rest. Unfortunately Popper sees the issue of realism almost exclusively in terms of this rather narrow school. As noted above, he reduces problems of knowledge and reality to the issue between 'positivism' and 'realism' and he distorts ideas until they fit this pattern. His treatment of Mach is a case in point.

Mach, according to Popper, is a 'positivist', the defender of a 'form of idealism' (R, p. 92) , who 'thought that only our sensations were real' (S, p. 18; R, p. 91) and rejected atoms for precisely that reason (R, p. 105). Now as regards atoms Mach asserted (A) that the atoms discussed in the kinetic theory of his time were untestable in principle, (B) that things untestable in principle should not be used in science but (C) that there was no objection to regarding them as '*provisional* aids' on the way to a 'more natural point of view' (references for the quotations in connection with Mach are given in chapter 7 below, as well as in chapters 5 and 6 of Vol. 2 of my *Philosophical Papers*).

(A) is a historical assumption. It was accepted by Einstein, who tried to establish the contact between atoms and observation which was missing at the time. (B) and (C) are cornerstones of Popper's philosophy which restricts science to testability but encourages speculation to go beyond it. Thus Mach did not 'reject [atoms] out of hand' (R, p. 191) – he accepted the idea, noted its untestable character and suggested that one look for something better. (He also decried the absurdity of metaphysical atomism which tried to 'explain sensations by motions of atoms'.) The path suggested by Mach was entered upon by Gibbs and Einstein (and had already been taken by Hertz, in his account of Maxwell's equations). In his early papers on statistical phenomena, Einstein criticized the kinetic theory because it 'had not been able to provide an adequate foundation for the general theory of heat' (*Ann. Phys.* (1902), p. 417): he tried to free the discussion of heat phenomena from special mechanical models and showed that some very general properties (first order differential equations for the temporal variation of state variables, which Mach had regarded as an important *empirical* fact; a unique integral of motion; and an analogue to Liouville's theorem) sufficed to obtain the desired results. Einstein's preference for 'theories of principle' over 'constructive theories' (theories tied to mechanical models), which guided him on the way to the special theory of relativity, was entirely in the spirit of Mach.

With regard to Mach's 'positivism', however, the situation is simple: it does not exist. 'Elements' are sensations – but only in certain contexts; 'they are simultaneously physical objects, viz. insofar as we consider other functional dependencies.' Sensation talk is not certain, it is based on a 'one sided theory' that must be supplemented by physiological research. Popper noticed some of the differences between Mach and the Mach-mythology — but he chose to disregard them. Using a procedure he severely criticises in others (cf. his objection to Kuhn in R, p. xxxiv), he says that Mach 'may legitimately be treated' as a sense-datist (R, p. 91 formulated in analogy to R, p. xxxiv). But the differences are even greater than described so far.

Mach objects to the 'limping', 'patchy' and 'uncertain' ways of induction. He objects to calling the natural sciences inductive sciences. He urges the scientist to rely on his 'instinct', to introduce 'principles' of great generality, to make 'bold intellectual moves' in order to 'reach a wider view', and to use such a view for subsuming and correcting particular results, results of precise experiments included (how does this square with Popper's claim that 'positivism from Berkeley to Mach has always opposed . . . speculation': Q, p. 172?). And while Einstein in some of his more philosophical writings started the process of knowledge from 'immediate sense experience' and emphasized the 'essentially fictitious' character of far reaching conjectures (he was an instrumentalist, in Popper's terminology, though an inconsistent one), Ernst Mach pointed out that 'not only mankind, but the individual, too, finds . . . a complete world view to whose construction he has made no conscious contribution – here everyone must begin' (compare this with Popper's 'we move, from the very start, in the field of intersubjectivity': R, p. 87). Mach regarded general features of the world not as 'fictions' but as 'facts', i.e. as real. As a matter of fact, we can say that Mach — who was a historian of science, and who did not 'because of the pressure of urgent work [rely] mainly on [his] memory' (R, p. xxvi: what urgent work? work of the same quality Popper is offering us now?) — was a much better critical rationalist than Popper can ever aspire to be; he did not stop short at dogmatic and inane declarations about reality (cf. R, pp. 83f.), he decided to *examine* the matter.

Similar remarks apply to Popper's treatment of Bellarmino. Why did Bellarmino (in his letter to Foscarini) suggest an 'instrumentalist' interpretation of the Copernican view? The reason was neither Aristotelian dogmatism nor a naive adherence to Bible passages (as Popper suggests). Jesuit astronomers had confirmed and improved upon Galileo's observations of the moon, Venus, and the moons of Jupiter, and the Ptolemaic system had been replaced by that of Tycho to account for the new phenomena. Physics and astronomy had been allowed on earlier occasions to change the interpretation of bible passages (for example, the spherical earth was a matter of course by the 11th century).

St. Bellarmino accepted that good arguments might similarly change established views about the motion of the earth. But, he added, there were no such arguments, and the faith which was an important part of the lives of common people should not be endangered by mere surmises. He was right on both points. The first point is now accepted by all serious students of the matter (the year was 1615). The second point hardly gets a hearing today because it is taken for granted that the ravings of specialists determine, but are never determined by, public concerns. But even Popper long ago warned us that social experiments should be made carefully, and in a piecemeal manner. Changing basic beliefs tied to powerful customs and familiar institutions, or 'opening minds', is a social experiment. It is a dangerous experiment, for opening minds in some respects always means closing them in others. Hence ideas with little support should not be introduced in an aggressive manner, they should be checked for their consequences and strengthened only when these are acceptable and better arguments become available. St. Bellarmino made precisely this point – to no avail.

This is how the naive alternative realism-positivism turns the history of science and civilization from a colourful interplay of fascinating and complex characters into a dreary exchange between 'the world's most distinguished living philosopher' (Martin Gardner on the cover of Popper's *Postscript*) and a collection of capable (Mach: a 'pathbreaking philosopher of nature', S, p. 135; Bohr: 'basically a realist', Q, p. 9), well-meaning (Hume, Mill, Russell: 'practical and realistic' in 'intention', R, p. 81; Heisenberg: 'understandable attitude', Q, p. 9), wonderful (Bohr: 'the most wonderful person I have ever met; Q, p. 9) but sadly confused interlocutors in urgent need of Popperian Enlightenment. The trivialization of history culminates in Popper's discussion of the quantum theory to which I now turn.

3 Quantum Theory

Towards the end of his *Postscript*, Popper sketches a cosmology that contains *change* and is *indeterministic*: 'cosmological fact[s]' (Q, p. 181) that 'correspond . . . closely to the commonsense view of the world' (p. 159). Catalogues giving weights to all possible states with laws for their development (p. 187) and conservation laws are invoked in support of this view. The conservation laws guide individual particles in a deterministic way, they remain valid for interactions, but they no longer 'suffice for determinism' (p. 190): we have *fields* of propensities for the appearance of *particles*. The old dualism, often commented upon by Einstein and Bohr, between fields and particles of equal reality is transformed into an Aristotelian dualism in which fields of *potentialities* become *actualised* as particles. To make the new dualism plausible Popper uses Dirac's hole theory of the positron: a positron is not a lump of matter; it is a possibility of occupation which may become actual as a result of interactions.

The theory has points of contact with the so-called S-matrix theory, especially in the interpretation given it by Chew. Both theories avoid reducing complex systems to smaller and smaller units until 'final building stones' (quarks, gluons, or what have you) are reached. Both object to regarding individual particles as 'given' and try to obtain their properties from interactions. Both accept a 'nuclear democracy' – no particle is more fundamental than any other particle. The formal principles of S-matrix theory—relativistic invariance, unitarity (sum of probabilities for all possible processes = 1) and analyticity (which is related to determinism for weights) — fit well into Popper's scheme, while Chew's 'bootstrap hypothesis' – that the basic field (the S-matrix in Chew's formalism) uniquely determines the properties of all particles (of all hadrons, in the present stage of development) — might play a role in a more 'scientific' version of it.

In addition, the two theories imply that the formalism of

quantum mechanics cannot remain unchanged. Popper, however, in the rest of his third volume, insists that he has an interpretation of the existing theory as well, that this interpretation is superior to the ideas of the theory's inventors, and that the difference is due to 'simple mistakes', 'muddles', and 'oversights' on their part. Not content with having found an interesting cosmology, he wants to show that other views are not worth considering. And so he turns from a capable representative of the Aristotelian tradition in metaphysics (Q, p. 165, p. 206) into a badly-informed, superficial and ill-tempered critic of physics.

Consider, for example, his 'end of the road thesis', the 'belief', that is, that 'quantum mechanics is final and complete' (Q, p. 5). This belief, says Popper, hindered research; for example, it created a resistance towards particles beyond the proton and the electron. Physicists remember things differently. Silvan Schweber, in a round table discussion on the history of particle physics (published in L. Brown and L. Hoddeson, eds., *The Birth of Particle Physics*, Cambridge University Press 1983, p. 265), commented on the 'dichotomy' between 'the revolutionary stand of the field theorists of the thirties as compared with the conservative stand of the post-World War II generation' (he mentions Bohr's well known refrain 'this is not crazy enough'), and likewise between this revolutionary attitude and 'an unwillingness to accept new particles'. Dirac, at the same conference, gave a reason for the second conservatism, (op. cit., p. 52): 'there were only two particles, two basic charged particles – electrons and protons. There were just two kinds of electricity, positive and negative, and one needed one particle for each kind of electricity.' Hanson, whose book on the positron Popper calls 'an excellent book' (R, p. xxix) that should be read by everybody (Q, p. 12), makes the same point. Conclusion: early particle physics was not inhibited by a quantum theoretical conservatism because (a) there was no such conservatism at the time and (b) the aversion to particle proliferation had a non-quantum theoretical source.

Popper makes the 'end of the road thesis' a special disease of quantum mechanics and connects it with the assertion that quantum mechanics is 'complete' (Q, p. 11). But quantum theoreticians were not the only people to think that they had arrived at final formulations: end-of-the-road statements are found in all branches of learning, even in relativity, even in Einstein. Popper himself calls some falsifications 'as "definitive" as general human fallibility permits' (R, p. xxiii). And 'completeness' as understood by Bohr and von Neumann means non-existence of variables *within* the uncertainty relations and *not* non-existence of further particles *obeying* uncertainty relations. The neutron and the positron were not '(previously) hidden variables' (p. 11) and no physicist ever thought they were. I pointed this out to Popper in 1962, during a conference at the Minnesota Center for the Philosophy of Science. His reaction: he blurred the notion of completeness. Now he does the same. 'The term', he says, 'has been used in several senses during this discussion' (p. 7): by Popper, yes – not by anybody else.

Discussing the narrower sense, Popper proceeds in the usual way. Having paid lip service to the wonderful personal qualities of Bohr (Q, p. 9), he uncovers two of his own most treasured weapons: misdescription and slander. He suggest that the Einstein-Podolsky-Rosen argument was rejected because of 'Bohr's authority, not counter argument' (Q, p. 149). But Einstein, certainly not a person to be impressed by authority, regarded Bohr's reply as an argument and added that it came 'nearest to doing justice to the problem' (Einstein's Schilpp volume, p. 681).

Popper asserts that 'the reply to Einstein and his collaborators consists of a surreptitious change of the theory which Einstein attacked, of a shifting of the ground' (p. 150). What he means is that while before the Einstein-Podolsky-Rosen argument (EPR) the uncertainties had been explained by an interaction, they were now explained in a different way. But the interaction view was held not by Bohr but by Heisenberg, and Bohr had commented on its unsatisfactory character long before EPR (cf. the 'addition in proof' to Heisenberg's 'Über den anschaulichen Inhalt der quantentheoretischen Kinematik und Mechanik', published in 1927, and section 3 of Bohr's 'The Quantum Postulate and the Recent Development of Atomic Theory' (1928); this difference between Bohr and Heisenberg is one of the reasons why it is historical nonsense to merge Bohr, Heisenberg, Pauli and others into a 'Copenhagen school' and then to attack this fictitious entity).

According to Popper, Bohr's 'changed' theory was 'much more harmless' (p. 150) than the interaction view: 'There was nothing more in it than that, at times, one coordinate system is applicable and at other times another coordinate system, but never both together. This', Popper concludes his verdict, 'leaves completely open what the particle itself does.' Of course it does, but what Popper has described so far is not Bohr's view but '[his] own version of' Bohr's view, as says himself (p. 150). So let us add what is explicitly stated in Bohr's reply to Einstein, namely that dynamical magnitudes such as position and momentum depend on the reference systems in a way that prevents their joint use: choose one reference system, and notions involving location cease to be applicable; choose another, and the same happens to notions dealing with motion; chose a third, and both are applicable only to a certain degree, to be determined by the uncertainty relations. Bohr compared the dependence just described with the relativistic dependence of all dynamical magnitudes on the reference system. Adding this assumption to Popper's 'version' of Bohr we arrive at a situation that no longer 'leaves it open what the particle itself does'. One can ridicule the assumption – and I am sure Popper would have done just that had he found it – but one cannot criticize Bohr for holding a view that does not contain it.

In a paper written twenty years ago, I explained Bohr's philosophy and defended it against a variety of criticisms, including criticisms which Popper had just published (and which are reprinted, with minor changes, in Q, pp. 35-85). I summarized as follows:

> Popper's criticism of the Copenhagen Interpretation, and especially of Bohr's ideas, is irrelevant, and his own interpretation is inadequate. The criticism is irrelevant as it neglects certain important facts, arguments, hypotheses and procedures which are necessary for a proper evaluation of complementarity and because it accuses its defenders of 'mistakes', 'muddles' and 'grave errors', which not only have not been committed, but against which Bohr and Heisenberg have issued quite explicit warnings. His own positive view [the view presented in the paper I criticized and reprinted as Q, pp. 35-85, and not the view of the Epilogue] . . . is a big and unfortunate step back from what had already been achieved in 1927.

There is not a line that needs changing in this summary and in the arguments that precede it. But it is interesting to see Popper's reaction. He mentions the paper in Q, p. 71, footnote 63 (added 1980). As is his habit, he disregards the critical remarks it contains and introduces an entirely fictitious complaint. He accuses Jammer (who had reported my criticism and had apparently agreed with it: see his *The Philosophy of Quantum Mechanics*, John Wiley & Sons 1974, p. 450) and, by implication, Bunge and myself (1) of having turned him into a subjectivist and

(2) of having equated his view with Bohr's. And he explains our alleged crime by our negligence – we took 'an almost accidental formulation', viz. Popper's '*experimental* setups', as proof of his subjectivism. But Jammer, who states very clearly (my view of) the difference between Bohr and Popper (ibid., p. 450, lines 12ff.), neither asserts nor implies that Popper failed to 'exorcise the observer from quantum mechanics' (Q, p. 35), while my point was that there is no observer to be exorcised because Bohr's probabilities are objective properties of either experimental arrangements, or natural situations (according to Bohr, Schroedinger's cat dies or stays alive even if there is no one around to look at it.) And as Bohr introduced his view much earlier I concluded that Popper, on probabilities, simply repeats Bohr. Popper, who seems to have read neither my paper nor Jammer's summary, again calls Bohr a subjectivist and infers that we (Jammer and I) made him, Popper, a subjectivist too. And why, according to Popper, is Bohr a subjectivist? Because of subjectivistic-sounding phrases whose context clearly reveals their objective content. Thus it is Popper who commits the crime (against Bohr) he accuses us (Jammer and me) of having perpetrated against him – a nice example of the quality of Popper's arguments. I conclude that Bohr's ideas are not even touched by Popper's 'analysis' – and this is unfortunate for Popper could have learned from Bohr, as others did before him, how to deal with the difficulties of realism caused by its remaining classical ingredients.

According to Popper (R, p. 149f), it is not easy to give an overall account of the structure of the world and the place of laws in it. Newton, who did not believe in action at a distance, explained both by making space the sensorium of god. The problem remains in a relativistic universe for here we again have features such as the 'absolute constancy of the electronic charge and mass; or, more generally, the absolute qualitative and quantitative identity of the properties of the elementary particles' (p. 151). Popper rejects idealistic solutions where the mind imposes its structure on the world. He concludes that 'we realists have to live with the difficulty' (p. 157). But the generalization ('we' realists) is entirely unwarranted.

Realists not satisfied with simpleminded notions of reality and not impeded by the idea that positivism is their only alternative have made some very interesting suggestions. To take an example (D. Bohm, *Wholeness and the Implicate Order*, London 1980, p. 145), let us compare the world with a photographic plate containing moving patterns for holograms and observations with methods of projection for obtaining holograms from it. A particular method of projection (a particular experimental arrangement) applied to a special part of the plate (world) gives rise to a hologram (experimental results) that mirrors the entire plate (world), though in an incomplete and confused way. A different method applied to a different part produces a different hologram which again mirrors the world in an incomplete and confused way. So much for the *physics* of the situation: it is 'objective' (if one wants to use such a superficial term) and it does not need an 'observer' to bring it about. Now we add some *historical* considerations (they played a large role in Bohr's philosophy): physicists working in the first domain try to explain its features. They find a theory that survives severe tests and seems to describe basic features of the world. The same happens in the second domain, but with a different theory and different concepts. We may try to reduce one theory to the other or to subsume both under a 'deeper' theory and so get closer to reality. The present model suggests a different point of view. It suggests that though we may improve our knowledge of a particular hologram (of the facts in a particular region of the world) we shall never get a complete view or an estimate of our distance from it: the phrase 'getting closer to the truth' makes no sense. We may also reduce a partial account of a particular hologram to a more complete account – but again it makes no sense to try to reduce a theory connected with one method of projection to a theory connected with a different method. To take an example: it makes sense to improve phenomenological thermodynamics and it makes sense to improve point mechanics. It makes no sense trying to reduce the one to the other. Why? Because it makes no sense to speak of the temperature of a system all of whose elements have precise positions. This physical/historical situation is precisely what Bohr had in mind when speaking of complementary aspects of the world. Bohm adds the idea of a substratum which can be explored by different kinds of experimental arrangements (means of projection) giving rise to different holograms but which is itself 'undefinable and immeasurable' (Bohm, p. 51 – italics in the original). Given the model the idea is most plausible. It gives us a glimpse of the many possibilities that lie beyond Popper's narrow horizon.

One final remark. In his attack on Bohr, Popper compares his own 'critical' philosophy with the alleged dogmatism of the Copenhagen Circle; more especially, he opposes his own 'arguments' to Bohr's irrationalism. This is a travesty of the facts. It is hard to find a group as aggressive and disrespectful of its 'leader' as the group of scientists, philosophers, students and Nobel Prize winners that regularly assembled around Bohr; and it is hard to find a thinker as aware of the many problems connected with our attempts to grasp reality as was Bohr. On the other hand, it is hard to duplicate the vapid servility that characterizes the Popperian Circle and it is almost impossible to dismantle all the myths, distortions, slanders and historical fairytales spread by its leader. A simple comparison between the editorial styles of Popper's *Postscript* and of Bohr's collected writings shows the enormous distance between friendly and occasionally mocking respect and tailwagging admiration. Bohr's ideas are food for thought for generations to come. Popper's 'ideas' are best forgotten as quickly as possible.

4 Historical Conclusion

According to Popper, 'our standards of rational discussion have seriously deteriorated since [the time of Boltzmann]. The decline started with the First World War and with the growth of the technological and instrumental attitude towards science' (Q, p. 157). There is now a 'general anti-rationalistic atmosphere which has become a major menace of our time' (p. 156).

There is a grain of truth in this complaint – but let us determine where it lies.

It does not lie in the area of the interaction of cultures. For though the main phenomenon here is still the relentless expansion of Western civilization, we have some first and very hopeful signs of a more respectful attitude towards ways of life different from our own. This respect is not merely a matter of sentiment, it has a practical foundation. It is connected with a series of interesting and surprising discoveries ranging from the discovery, towards the end of the last century, of the magnificent late paleolithic art up to the recent discovery, or rather rediscovery, of the efficiency of non-Western medical systems. The knowledge possessed by non-Western civilizations and by so-called primitive tribes is truly astounding. It aids its practitioners in their own social and geographical conditions *and* contains elements that exceed what the corresponding elements of Western civilization can do for us. As the discoveries become more widely known, blind admiration for Western science and for the 'rationalism' that goes with it gives way to a more differentiated and, I would add, more humanitarian attitude: *all cultures* and not only the cultures connected with Western science and rationalism have made and, despite great obstacles, are continuing to make contributions from which humanity as a whole can benefit.

This attitude is not new – it has great and important ancestors. It characterized the 'First Internationalism' of the civilization of the late Bronze Age in the Near East. The peoples of this period constantly fought each other – but they exchanged languages, works of art and literature, styles, technologies, minerals, grain, artists, generals, prostitutes and even gods. The attitude was resurrected and defended with vigour by the sophists; it was the basis of Herodotus's wonderful history, as we have seen. It was the philosophy of Montaigne and of his followers before and during the Enlightenment. It was then pushed aside by the strident expansion of a scientistic philosophy. Its return in our own century means that people at last look at strange things in a more reasonable way. This increase of 'rationality', incidentally, occurs in an area of far greater importance than Western theoretical cosmology, Popper's main measure of excellence (cf. his remarks on Newton, S, p. 208, and Einstein, Q, p. 158, which show a characteristic narrowness of vision). Needless to say, our rationalists are not at all pleased: they mumble darkly of 'relativism' and 'irrationalism' – new inarticulate substitutes for the old and never abandoned curse: *anathema sit!*

The deterioration of standards of rationality deplored by Popper cannot be located in physics either (though there are of course idiots here as anywhere else). On the contrary, the rise of new forms of organization (CERN, for example) has led to a situation where scientists become craftsmen, speculators, administrators just like Giotto, Brunelleschi, Ghiberti and other Renaissance artists. Ethical problems impose themselves with a force entirely unknown at the time of Boltzmann.[1] Bell worked at CERN (I think he is still there), most field theoreticians had at least a passing acquaintance with Los Alamos, many of them have strong interests in other fields (read Feynman's and Dyson's autobiographies as well as Dyson's book on the nuclear threat) — and all this without an increase of 'instrumentalism'. Why should experimentalists so eagerly look for isolated quarks or magnetic monopoles, why should they attempt to catch neutrinos from the centre of the sun to explore the solar energy household, if all these things were nothing but instruments? I have before me a volume edited by J.A. Wheeler and W.H. Zurek containing articles on the interpretation of quantum mechanics (*Quantum Theory and Measurement*, Princeton 1983). Popper and/or his editor must have liked it, for they hurried to get into its bibliography (which contains a reference to proof copies of the *Postscript*). There are papers ranging from the mind-body problem via experimental tests of locality to purely formal considerations. Looking at this material I notice no deterioration of standards – on the contrary, compared with 'the controversies around Boltzmann' (Q, p. 157) we have a considerable refinement of argumentation accompanied by vastly increased philosophical depth. And let us not forget the work of Bohr and Heisenberg, their real work, that is, and not Popper's caricatures. There is only one area where the deterioration is obvious: Popper's own field, the philosophy of science — and Popper has done his best to keep it that way. So, let me now make a few concluding remarks on *this* situation.

Late nineteenth century philosophy of science was developed by scientists in close connection with their work. It was pluralistic and it removed conditions that had been regarded as essential for knowledge. Of course, every author favoured some procedures and rejected others – but most scientists agreed that such personal preferences should not be turned into 'objective' boundaries of research. 'The best means of promoting the development of the sciences', wrote the physical chemist, historian and philosopher of science Pierre Duhem after a vigorous diatribe against model building, 'is to permit each form of intellect to develop itself by following its own laws and realising fully its type' (*The Aim and Structure of Physical Theory*, New York 1962, p. 99). 'I must confess', wrote von Helmholtz, perhaps the most versatile scientist of the 19th century (Preface to H. Hertz, *Die Principien der Mechanik*, Leipzig 1894, p. 21, quoted by Duhem), 'that so far I have retained the latter procedure [mathematical equations instead of models] and have felt safe with it – but I would not like to raise general objections against the way which such excellent physicists . . . have chosen.' 'Discovery', Duhem points out (op. cit., p. 98) 'is not subject to any fixed rule. There is no doctrine so foolish that it may not some day be able to give birth to a new and happy idea. Judicial astrology has played

---

1. For an excellent presentation of a particular phase of this development, see R. Rhodes, *The Making of the Atomic Bomb*, New York 1986.

its part in the development of the principles of celestial mechanics.' And Ludwig Boltzmann concluded an interesting survey of new ideas and methods in theoretical physics by saying: 'It was a mistake to regard the older procedures as the only correct ones. But it would be equally one sided to reject them now completely, after they have led to so many important results . . . ' ('Über die Methoden der Theoretischen Physik', *Populäre Schriften*, Leipzig 1905, p. 10).

The pluralism implicit in these quotations found support in Darwin's theory. Before Darwin it had been customary to view organisms as divinely created and therefore perfect solutions to the problem of survival. Darwin drew attention to numerous 'mistakes': life is not a carefully planned and meticulously performed realization of clear and stable aims; it is unreasonable, wasteful, it produces an immense variety of forms and leaves it to the particular stage it has reached (and the natural surroundings existing at the time) to define and eliminate the failures. Similarly, so Mach, Boltzmann and other followers of Darwin inferred, the development of knowledge is not a well planned and smoothly running process; it, too, is wasteful and full of mistakes; it, too, needs many ideas and procedures to keep it going. Laws, theories, basic patterns of thinking, facts, even the most elementary logical principles are transitory results, not defining properties of this process. Scientists, accordingly, are not obedient slaves who on entering the Temple of Science anxiously try to conform to its rules; they do not ask 'what is science?' or 'what is knowledge?' or 'how does a good scientist proceed?' and then adapt their research to the limitations contained in the answer; they forge ahead and constantly redefine science (and knowledge, and logic) by their work.

The view just described makes history an important part of scientific research. According to Ernst Mach (*Erkenntnis und Irrtum*, Leipzig 1917, p. 200), 'the schemata of formal logic and of inductive logic are of little use [to the scientist] for the intellectual situation is never exactly the same.' To understand science, says Mach, is to understand the achievements of great scientists. Such achievements are 'very instructive' not because they contain common elements which the researcher must detach and learn by heart if he wants to become a good scientist, but because they provide a rich and varied playground for his imagination. Entering the playground 'like an attentive wanderer' ('*wie ein aufmerksamer Spaziergänger*': op. cit., p. 18), the researcher develops his imagination, makes it nimble and versatile and capable of reacting to new challenges in new ways. Research, accordingly, 'cannot be taught' (p. 200), it is not 'a bag of lawyer's tricks' (p. 402, footnote), it is an *art* whose explicit features reveal only a tiny part of its possibilities and whose rules are often suspended and changed by accidents and/or human ingenuity. Many nineteenth-century scientists thought along similar lines, as we have seen. And this was not a philosophical luxury without effect on the practice of science; this brought about the two most fascinating theories of twentieth-century physics: the quantum theory and the theory of relativity. Their creators were fully aware of the connection.

Thus Niels Bohr pointed out that 'in dealing with the task of bringing order into an entirely new field of experience, we [can] hardly trust in any accustomed principles, however broad' (*Albert Einstein, Philosopher-Scientist*, ed. P.A. Schilpp, Evanston 1949, p. 228), while Leon Rosenfeld added that 'in speculating about the prospects of some line of investigation [Bohr] would dismiss the usual considerations of simplicity, elegance or even consistency' (*Niels Bohr. His Life and Work as Seen by His Friends and Colleagues*, ed. S. Rosental, New York 1967, p. 117). The best account is Einstein's. Commenting on the efforts of 'rational' and 'systematic' philosophers, he wrote (Schilpp Volume, p. 684):

> No sooner has the epistemologist, who is seeking a clear system, fought his way through such a system, than he is inclined to interpret the thought content of science in the sense of his system, and to reject whatever does not fit into his system. The scientist, however, cannot afford to carry his striving for epistemological systematicity that far . . . ; the external conditions which are set for him by the facts of experience, do not permit him to let himself be too much restricted in the construction of his conceptual world by the adherence to an epistemological system. He therefore must appear to the systematic epistemologist as a type of unscrupulous opportunist.

Now it is surprising to see how little effect such ideas have had on philosophy, the social sciences and intellectuals in general. Even worse, neopositivism, which arose while the revolution of modern physics was in full swing, used the name of science to propagate a rigid, narrow-minded and unrealistic point of view. Neopositivism was not a bold and progressive reform of philosophy; it was a descent into a new philosophical primitivism. Surrounded by fundamental changes in physics, biology, psychology and anthropology, by interesting and much-debated points of view in the arts, and by unforeseen developments in politics, the fathers of the Vienna Circle withdrew to a narrow and badly constructed bastion. The connections with history were severed; the close collaboration between scientific thought and philosophical speculation came to an end; terminology alien to the sciences and problems without scientific relevance took over and the image of science was distorted beyond recognition. Fleck, Polanyi and then Kuhn compared the resulting ideology with its alleged object – science – and showed its illusionary character. Their work did not improve matters. Philosophers did not return to history. They did not abandon the logical charades that are their trademark. They enriched them with further empty gestures, most of them taken from Kuhn ('paradigm', 'crisis', 'revolution') without regard for context, and thus complicated their doctrine without bringing it closer to science. Pre-Kuhnian positivism was infantile, but relatively clear. Post-Kuhnian positivism has remained infantile – but it is also very unclear. Where is Popper situated in this mess?

He started with a technical suggestion that remained within the framework of positivism: separate the problem of demarcation from the problem of induction, solve the first by falsifiability and the second by a method of bold conjectures and severe tests. The suggestion was technical because it was formulated in the logical terminology favoured by positivists and because it followed positivism in replacing real scientific theories by logical caricatures (see above, and cf. also Popper's repeated assertion that his theory of science is not a historical theory and cannot be criticized by historical evidence). Popper's contribution was to confirmation theory, not to scientific practice. Popper then incorporated this technical suggestion into a wider view, critical rationalism, and tried to illustrate it by historical episodes: the battle about falsification, he wanted to say, was not just positivistic backbiting – it had historical scope. This was correct in one sense, incorrect in another. Popper repeats what others said before him, but he repeats it badly and without the historical perspective of his forerunners.[2]

Still, some nervous scientists who had taken positivism seriously and who read Popper were much relieved; they could now speculate without having to tremble for their reputations. This explains part of Popper's popularity — a popularity sometimes transferred to his other views, which were also liked for themselves, partly because they were simple, partly because they built a philosophical altar for science. But the simplicity offered by Popper is the result not of penetration but of simpleminded-ness, and those who praise his physics (Bondi, Denbigh, Mar-genau and others) have much in common with the early oppon-ents of Einstein who praised Lenard and Stark 'because they were unable to follow the difficult ways of thinking modern physics demanded from them' (Elisabeth Heisenberg, *Das Poli-tische Leben eines Unpolitischen*, Munich 1982, p.36). Inde-pendent scientists never needed a simplifier, a methodological crutch or an altar; and the keys to freedom which Popper offered to the more cowardly members of their craft are and always were in their possession (cf. the above sketch of the philosophy of nineteenth-century scientists, of Einstein and of Bohr). There is absolutely no need to pay a price for this freedom and to exchange one slavery (positivistic puritanism) for another (Pop-perian pidgin science).

2. This is realised even by some of Popper's fans. Thus P. Medawar, *Advice to a Young Scientist*, Harper and Row, New York 1979, pp. 90f, writes that 'Whewell first propounded a view of science of the same general kind as that which Karl Popper has developed into a thoroughgoing system' – i.e. Popper = Whewell petrified. O. Neurath (letter to R. Cainap, Oxford, Dec 22 Frankfurt 1987, 327) hits the nail on the head: 'I read Popper again. I hope that after so many years you will see, how empty all that stuff is . . . What a decrease after Duhem, Mach, etc. No feeling for scientific research.' Lakatos, late in his life, came to the same conclusion.

7

Mach's Theory of Research and its Relation to Einstein

Introduction

In his autobiographical notes, Einstein credits Mach with having shaken the dogmatic faith in the fundamental role of mechanics. Mach's *Geschichte der Mechanik*, writes Einstein,[1] 'exercised a profound influence upon me in this regard while I was a student. I see Mach's greatness in his incorruptible scepticism and independence; in my younger years, however, Mach's epistemological position also influenced me greatly – a position which today appears to me essentially untenable.'

According to this passage Mach was engaged in two kinds of activity. He criticized the physics of his time and he also developed an 'epistemological position'. The two activities seem to have been relatively independent of each other; for Einstein, in his later years, accepted the one and rejected the other. He also described them differently. Mach the epistemologist, he said,[2] regarded 'sensations as the building blocks of the real world', while Mach the physicist criticized absolute space without ever leaving the domain of physics.[3]

In the following essay I shall try to separate Mach's physical arguments from his 'epistemology'. It will turn out that a separ-

---

† Dedicated to Adolf Grünbaum on the occasion of his sixtieth birthday.

1. P.A. Schilpp ed. *Albert Einstein, Philosopher-Scientist*, Evanston 1949, p. 20.

2. Letter to Michele Besso of 6 January 1948, quoted from G. Holton, *Thematic Origins of Scientific Thought*, Cambridge 1973, p. 231.

3. Cf. Einstein's account in Schilpp, ed., op. cit., p. 28.

ation is not difficult to achieve. Mach's physical arguments, taken together, constitute a philosophy of science which differs from positivism, is in line with Einstein's research practice (and some of Einstein's more general observations on research) and entails entirely reasonable objections against nineteenth century atoms and the special theory of relativity. We shall also see that where Mach and Einstein differed it was Einstein who talked positivism while Mach gave a much more complex account of scientific and commonsense knowledge. Mach's 'epistemology', however, turns out to be no epistemology at all. It is a general scientific theory (or theory – sketch) comparable in form (though not in content) to atomism, and different from any positivistic ontology.

1 Mach on the Use of Principles in Research

In chapter 4, section 3 above, I explained how Ernst Mach, using Simon Stevin's thought experiment as an illustration, argued for intuitively plausible principles and against a stepwise inductive approach (*Mechanik*, chapter 1, section 2[4]). Proceeding in this way, he said, is 'not a mistake. If it were a mistake, then all of us would share it. It is, moreover, certain that only the strongest instinct combined with the strongest conceptual power can make a person a great scientist' (p. 27; cf.E 163). Indeed, 'one can say that the most important and the weightiest extensions of science are made in this manner. This procedure, practiced by great scientists, of harmonizing particular ideas with the general outline (*Allgemeinbild*) of a domain of phenomena, this constant regard for the whole when contemplating individual effects can be called a truly philosophical procedure' (p. 29).

The procedure affects our concepts. Principles disregard the peculiarities of concrete physical events. Basing science on principles forces us to free such events 'from disturbing circumstances' (p. 30) and to present them in an idealised form: edges and beams are replaced by inclined planes and levers[5] just as edges and polished surfaces were replaced by lines and planes in

4. I am quoting from the 9th edition, Leipzig 1933. Numbers in brackets are pages of the book. Numbers preceded by E are pages of *Erkenntnis und Irrtum*, Leipzig 1917. Original emphasis, unless otherwise indicated.

5. Stevin himself considers a thread with fourteen equally heavy and equally distant balls – *cf*. the illustration from his book which appears on p. 31 of the *Mechanik*.

geometry (p. 30): we 'actively reconstruct the facts with the aid of exact concepts and [now] can master them in a scientific way' (p. 30).

2 Einstein's Use of Principles

Now consider Einstein's description of the way in which he arrived at the special theory of relativity.[6] Facing a difficult situation in physics he tried to 'discover true laws by means of constructive efforts based on known facts'; he 'despaired' of achieving success in this manner. Guided by the example of thermodynamics which starts with principles, not with facts, he became convinced 'that only the discovery of a universal principle would lead . . . to secure results.' He found a principle by means of the following thought experiment: 'if I pursue a beam of light with the velocity c (the velocity of light in a vacuum) I should observe such a beam of light as spatially oscillating electromagnetic field at rest. However, there seems to be no such thing, whether on the basis of experience, or according to Maxwell's equations.'

There is hardly any difference between this procedure and the moves described and recommended by Mach.

The similarity extends to details. Thus Einstein on more than one occasion denied having been influenced by the Michelson – Morley experiment: 'I guess I just took it for granted that it was true.'[7] 'For it does not matter if the experiment is really performed', wrote Mach in the case of Stevin (p.29), 'if only the success is beyond doubt.' Asked about the source of his conviction Einstein referred to intuition and 'the sense of the thing' (*die Vernunft der Sache*),[8] parallelling Mach's emphasis on the instinctive (intuitive) nature of fruitful principles. 'It fits the economy of thought and the aesthetics of science', said Mach (p. 72),

6. P.A. Schilpp ed., op. cit., p. 52.

7. R. S. Shankland, 'Conversations with Albert Einstein', *Am. J. Phys.* 31 (1963), p. 55: Cf. also Einstein's reaction to Eddington's cable of 1919 as reported in Ilse Rosenthal – Schneider's reminiscences (quoted from G. Holton, *Thematic Origins of Scientific Thought,* Cambridge 1973): 'But I knew that the theory is correct'. Concerning the equality of inertial and gravitational mass Einstein said: 'I had no doubt about its strict validity even without knowing the result of the admirable experiment of Eötvös which – if my memory is right – I only came to know later': *Ideas and Opinions,* New York 1954, p. 287.

directly to *recognize* a *principle* . . . as a key for understanding *all* facts of a domain and to *see* in ones mind how it penetrates *all* facts – rather than finding it necessary to prove it in a patchy and limping way, using propositions *accidentally* known to us as a foundation . . . Indeed, this eagerness to prove leads to a *false and misconceived rigour*: some statements are regarded as more secure and as the necessary and uncontestable foundation for others while they have only the same certainty, or even a lesser degree of it.

This, of course, means preferring Einstein to Lorentz, as becomes clear from Lorentz's description of his own procedure:[9]

Einstein simply postulated what we have deduced, with some difficulty and not altogether satisfactorily, from the fundamental equations of the electromagnetic field. By so doing he may certainly take credit for making us see in the negative results like those of Michelson, Raleigh and Bruce not *a fortuitous compensation of opposing effects*, but the manifestation of a general and fundamental principle.

According to Mach, principles are capable and in need of being tested by experience (p. 231). Einstein agrees. Science, he says, attempts to 'find a unifying theoretical system',[10] but he adds that 'the logical foundation is always in greater peril from new experiences or new knowledge than are the branch disciplines with their closer empirical contact. In the connection of the foundation with all the single parts lies its great significance, but likewise its greatest danger in face of any new factor.'[11] On the

8. Letter to Besso quoted in Carl Seelig, *Albert Einstein*, Zürich 1954, p. 195. Cf. also Born's observation of 4 May, 1952 and Einstein's reply of 12 May in the *Born – Einstein Letters* (New York 1971) – 'Only intuition resting on sympathetic experience can reach' the basic laws: address delivered at a celebration of Planck's sixtieth birthday (1918) before the Berlin Physical Society, quoted from *Ideas and Opinions*, p. 226.

9. *The Theory of Electrons*, New York 1952, p. 230, my emphasis. I do not assert that Mach himself preferred Einstein to Lorentz – there exists no evidence on this point. But the two ways outlined by Lorentz fit perfectly the two ways described by Mach – and Mach preferred the use of comprehensive principles, not the use of isolated facts and assumptions and tortuous derivations therefrom.

10. 'The Fundaments of Theoretical Physics', *Science* (1940) quoted from *Ideas and Opinions* p. 234. Cf. Mach: 'those ideas which can be retained in the largest domain and most extensively complement experience are the *most scientific* – p. 465.

11. Op. cit., p. 325. Mach regards instinctive general principles as more trustworthy than individual experimental results precisely because they are in potential conflict with a comprehensive domain of facts and have survived in spite of it. Cf. Section 5 below.

other hand he was not willing to abandon a plausible idea just because it conflicted with some experimental result, paralleling Mach's emphasis on the authority of instinctive principles and the need to adapt the empirical facts to them: there is no better way of describing Einstein's procedure in his relativity paper than by repeating, with the key elements exchanged, the brief account Mach gives of Stevin's argument.[12]

3 Some Criticisms of Mach Refuted

Consider now some popular comments on the relations between Mach and Einstein.

Professor Arthur Miller, who has written an excellent, perceptive and very detailed book on the prehistory and the early interpretation of the special theory of relativity,[13] wants to explain Mach's criticism of that theory in the foreword to his *Optics*. I don't think he describes the criticism correctly when saying that Mach 'bluntly repudiated the relativity theory' (Miller, p. 138).[14] Mach[15] promises to explain 'why and to what extent (*inwieferne*) he rejects relativity *in his own thinking (für mich)*',[16] which means that the question of the nature and the bluntness of the rejection is left undecided and that the answer is postponed to a future publication (which never appeared). Nor

12. Holton, 205ff., connects Einstein's unique style (starting from principles instead of experiments, or problems) with Föppl. He might as well have connected it with Mach whom Einstein had studied and whom Föppl revered. Arthur Miller writes (Holton and Elkana, eds., *Albert Einstein, Historical and Cultural Perspectives*, Princeton 1982, p. 18) that Einstein 'took recourse in the Neo-Kantian view that was predicated on the usefulness of organizing principles such as the second law of thermodynamics'. Considering Einstein's admiration for Hume which Miller mentions (loc. cit.), this is hardly likely. But the very same kind of principles which Neo-Kantians tried to establish in an a priori manner were discussed and recommended by Mach, who based them on instinct, explained why instinct should be relied upon (section 5 below) and rejected a priori arguments (p.73.) On a priori assumptions *cf.* also Einstein's letter to Born, undated, *The Born – Einstein Letters*, New York 1971, p. 7.

13. *Albert Einstein's Special Theory of Relativity*, Reading Massachusetts 1981.

14. Einstein himself interprets Mach in the same way: 'Mach rejected the special relativity theory *passionately*' – letter to Besso of 6 January 1948, quoted from Holton *op. cit.*, p. 232, my emphasis.

15. *Die Prinzipien der Physikalischen Optik*, Leipzig 1921.

16. The italicized restriction is omitted in the English translation, Dover publications, page viii.

can we accept Miller's reasons for the criticism.

According to Miller (p. 167), 'Einstein's a priori declaration of the postulates of relativity already indicated that he had gone beyond Mach.'

Now it is true that Einstein starts his 1905 paper not with experimental facts but with postulates and that he continues by deriving consequences from them – but this is precisely the procedure described and recommended by Mach. It is also true that Mach emphasized the need to test the principles by experience (p. 231): but here again we find agreement, as we have seen. Mach's further comment that the principles 'may be used as starting points for mathematical deductions' because of 'the stability of our surroundings' (p. 231) again brings us close to Einstein who regarded the special theory of relativity as being valid in special and stable surroundings only.

'The axiomatic status of Einstein's two postulates of relativity', writes Miller (p. 166), 'placed them outside the scope of direct experimental observation.' Correct – except for the implication that Mach would have disapproved. Even Miller's (correct) observation that 'data (in Einstein) could also mean the results of *Gedanken experiments*' (p. 166) does not lead to a conflict with Mach, as we have seen: the apparent clash between Mach and Einstein cannot have been a clash about the appropriate research procedure. [17]

Considering fundamental principles such as the first and second law of thermodynamics, Newton's first law, the constancy of light velocity, the validity of Maxwell's equations, and the equality of inertial and gravitational mass, Gerald Holton writes [18] that 'none of these would have been called "facts of experience" by Mach.'

Holton *asserts* that Mach does not apply the term 'facts (of experience)' to principles of a certain generality and he *implies* that Mach would have objected to using such principles as a basis of argument. Both the assertion and the implication conflict with important parts of Mach's work. As I have tried to show in sections 1 and 2 and as will emerge in section 4, Mach is very critical towards naive inductive procedures and prefers the direct and 'instinctive' use of principles of great generality. Moreover, there exist many passages in his work which use the term 'facts of

17. The same criticism applies to R. Itagaki's statement that 'the direction from principle to experiment is diametrically opposed to Mach's methodology': *Historia Scientiarum*, 22, 1982.

18. *Thematic Origins*, p. 229.

experience' in precisely the way denied to him by Holton.[19]

A further criticism, also mentioned in Holton's book,[20] comes from Einstein. According to Einstein, 'Mach's system studies the existing relations between data of experience: for Mach science is the totality of these relations. That point of view is wrong and in fact what Mach has done is to make a catalogue, not a system.' Many philosophers and historians have repeated the criticism. They can be refuted by noticing how often and how insistently Mach emphasizes the need to free general facts from the peculiarities of individual observations and experiments and to always 'pay attention to the whole' (p. 29). As he represents it, the historical development of mechanics consists in the gradual revelation, the 'step by step recognition' (p. 244), of, basically, 'one big fact'. The most productive scientists are those who, gifted with a 'wide view' (*Weitsichtigkeit* – E 442, and cf. 476) can 'clearly perceive principles through all the facts' (pp. 61, 72, 133, 266 and many other places), 'recognizing a principle directly as the key for the understanding of *all* facts in a domain and seeing in their minds how it penetrates *all* facts' (p. 72), '*intuiting* [it] in

19. Examples are: the principle of the parallelogram of forces (pp. 44ff), the law of inertia (p. 264, 244), the existence of masses (p. 244) whose magnitude is independent of the method (direct, or indirect) used for determining them (E 175 – this means accepting the equality of gravitational and inertial mass as a fact of experience). Also the view that what leads to motion determines accelerations no matter whether we are dealing with 'terrestrial gravity, the attraction of the planets [or] the action of magnets' (p. 187) is explicitly described as expressing 'one single big fact' (p. 244). 'I completely agree with Petzold', writes Mach at a different place (p. 371 – original all italics), 'when he says: "hence all statements made by Euler and Hamilton are nothing but analytic expressions of the *empirical fact* that the processes of nature are uniquely determined." ' Energy conservation is called an empirical fact both directly and by implication (pp. 437, 477ff.) though it was revealed by thought experiments, not by careful experimental research (E 194). Faraday is praised as a researcher who restricted physics 'to expressing what is factual' (p. 473 – this includes the idea of contact action, E 443), 'but physicists caught up in a physics of action at a distance started understanding his ideas only when Maxwell translated them into a language they were familiar with' (E 442): there is no doubt that Mach would have regarded 'the existence (and the behaviour) of electric and magnetic fields in a vacuum which arose from the work of Faraday, Maxwell and Hertz' (E 444) as an empirical fact. Even the parallel axiom is counted among the principles we accept intuitively (E 414) and use then as a basis for an 'exact reconstruction of facts' (p.30). Discussing the role of principles in research and especially in the construction of facts, he calls them 'observation(s) which are as legitimate as any other observation' (*so gut eine Beobachtung als jede andere*: p. 44, referring to the principle of the parallelogram of forces).

20. Holton, p. 239. The quotation is from Einstein's Paris lecture of 6 April, 1922.

the processes of nature' (p. 133: the reference is to Galileo; cf. also E 207), thus 'comprehending more in one glance' (p. 133) than naive observers who, having a 'more narrow view' (*Kurzsichtigkeit* – E 442), are detracted by 'secondary circumstances' (p. 70 – *Nebenumstände*; cf. E 414: accidental disturbances) and 'find it difficult to select and pay attention to what is essential' (p. 70). Productive scientists, accordingly, do not enumerate facts and arrange them in lists, they either 'reconstruct' them (p. 30) or engage in 'constructive efforts', building 'ideal cases' (E 190f.) from their 'own reservoir of ideas' (E 316). Nor are they content with consistency – they look for *'an even greater harmony'* (E 178, my emphasis) and they find it in the general facts and instinctive principles already described.

4 Mach on Induction, Sensations and the Progress of Science

Mach's views on science emerge very clearly from his attitude towards induction. 'It is really strange', he writes (E 312),

> that most scientists regard induction as the principal means of research as if the natural sciences had no other business than to arrange manifest individual facts directly in classes. The importance of this activity is not denied, but it does not exhaust the task of the scientist; above all, the scientist must find the relevant *characteristics* and their *connection* and this is much more difficult than classifying what is already known. There is therefore no justification for calling the natural sciences inductive sciences.

What are relevant characteristics and how are they found?

According to Mach the relevant characteristic of classical dynamics is that there are masses, that different ways of measuring masses always lead to the same result and that whatever induces motion (terrestrial gravitation; attraction of planets, of the moon, of the sun; magnetism; electricity) determines accelerations, not velocities (pp. 187, 244 etc.); in short, it is what is described by the *principles* of mechanics. We have already seen that instinct and intuition play an important role in the discovery of principles (E 315: *'intuition* is the basis of all knowledge'). As Mach says:

The psychological operation by means of which we gain new insight and which is frequently though quite unfittingly called induction is not a simple process – it is very complex. It is not a logical process though logical processes can be interpolated as intermediate and auxiliary links. *Abstraction* and *imagination* play a major role in the discovery of new knowledge. The fact that method can aid us very little in these matters explains the *air of mystery* (*das Mysteriöse*) which according to Whewell characterizes inductive findings. The scientist looks for an enlightening idea. To start with he knows neither the idea nor the manner in which it can be found. But when the aim and the path to it have revealed themselves, the scientist is at first as surprised by his findings as a person who, having been lost in a wood, suddenly, on leaving the thicket gains an open view and sees everything lying clearly before him. Method can impose order and improve results (*kann ordnend und feilend eingreifen*) only after the main thing has been found (E 318ff.).

Finding principles involves observations side by side with ingredients which the scientist 'adds *on his own*, using his own reservoir of ideas'. Thus Kepler's tentative assumption of the ellipticity of the orbit of Mars is his own construction.[21] The same applies to Galileo's assumption of the proportionality of velocity and time for free fall and Newton's proportionality of speed of cooling and temperature difference (E 316). The nature and the quality of the additional ingredients depend on the scientist, the shape of the science of his time, and the extent 'to which [he] is satisfied with the mere statement of a fact' (E 316). Newton's thinking, for example, is characterized by great boldness and considerable imagination and, 'indeed, we have no hesitation to regard the latter as the most important' element of his research (p. 181): 'grasping nature by *imagination* must precede *understanding* so that our concepts may have a lively and intuitive content' (E 107).[22]

21. Cf. E. 152. The construction omits perturbations and therefore builds 'ideal cases' (E 190ff.).

22. Here is a further similarity between Mach and Einstein. Einstein repeatedly emphasized that research cannot rest content with sensations and concepts ordering sensations but that it needs objects which are 'to a high degree independent of sense impressions' (*Ideas and Opinions*, p. 291). But while Einstein regarded these objects as 'arbitrary' and as 'free mental creations' (loc. cit. and many other passages), thus ceasing to ask questions at a decisive point, Mach examined their origin and the nature and source of their authority. As a result he gave arguments for upholding general points of view in the face of contrary facts while Einstein, who often and exuberantly violated the rules of a naive falsificationism, could appeal to his subjective convictions only. This matter will be discussed in greater detail in Section 5 below.

We have seen that abstraction, according to Mach, 'plays an important role in the discovery of knowledge' (E 318). Abstraction seems to be a negative procedure: real physical properties, colours (in the case of mechanics), temperature, friction, air resistance, planetary disturbances are *omitted*. For Mach this is the side effect of *positive* and *constructive* work that is 'added' (E 316) by the scientist and used by him to 'rebuild' facts. Abstraction, as interpreted by Mach, is therefore '*a bold intellectual move*' (*ein intellektuelles Wagnis*: E 140, and cf. E 315 on the connection between abstraction and attention). It can misfire, it 'is justified by *success*' (E 140). Mach's famous slogan 'science means adapting ideas to facts and to each other' (p. 478, and many other places) has to be read accordingly. Adapting ideas to facts does not mean *repeating* the *unchanged* facts in the medium of thought, it is a dialectical process that transforms both ingredients. Let us recall, this time in somewhat greater detail, how the process unfolds.

Trying to find order in the world, the scientist looks for principles and finds them either in a 'limping', 'patchy' and 'uncertain' way, by consulting experiments, or instinctively, with the aid of bold thought experiments and generalizations drawn therefrom. Principles define a style of thinking and invite us to 'sketch' (p. 73), or 'idealize' (p. 30; E 190f.), the known facts in this style, 'abstracting' from the elements not contained in it. This is a truly creative enterprise which connects facts and ideas by changing and rebuilding both.[23] The results are not unique.[24] Different principles suggesting different methods of abstraction idealize or 'sketch' facts in different and even contrary directions, emphasizing 'now this, now some other aspect of phenomena' (p. 73). Thus Black, viewing heat as a substance and therefore assuming the conservation of heat, introduced latent heat to account for freezing and evaporation while nineteenth-century thermodynamicists allowed for the change of heat into other forms of energy (E 175). Similarly Benedetti, assuming

23. 'It is *logically* possible that somebody produces a purely phoronomic analysis of Keplerian motion and hits upon the idea of describing it by accelerations which are in the inverse square to the radius from the sun and directed towards it. But this process, in my opinion, is *psychologically* unthinkable. Why should somebody without a physical conception to guide him, hit upon *accelerations*? Why not use the first or the third differential quotient? Why use, out of the infinitely many possible analyses of motion into two componenents the one which yields such a simple result? For me already the analysis of the parabolic trajectory of thrown objects is very difficult without the guiding idea of gravitational acceleration' (E 147ff).

impetus, also assumed a natural fading away of it while Galileo who later in his life connected the laws of inertia with relative motions could restrict himself to physically identifiable obstacles (p. 263): the adaptation of facts and ideas 'can proceed in many *different* ways' (E 175). Idealizations arising from different areas or from different principles in the same domain occasionally clash and give rise to paradoxes. (One example of such a clash is Einstein's thought experiment as described in Section 2.) Such paradoxes are 'the *strongest moving force* of research (E 176; cf.264 and *passim*). 'One can never say that the process has completely succeeded and that it has 'come to an end' (p. 73) and one can therefore never say that a fact – any fact – has been completely and exhaustively described. Even sensation talk 'involves a one sided theory' which must be tested and developed by research.[25]

According to Mach, 'the mental field' — the domain of thoughts, emotions, strivings etc. — 'cannot be fully explored by introspection. But introspection combined with physiological research which examines the physical connections can put this field clearly before us and thereby make us acquainted with our inner being'[26]: introspection does not suffice. The whole nature of mental events is revealed by an enterprise that contains introspective psychology and physiological research as mutually dependent research strategies.

Mach had two reasons for basing not only psychology but scientific research as a whole on such mixed strategies. The first reason was his critical attitude: he wanted to examine even the most general and the most firmly entrenched ingredients of

24. For Einstein the fact that experience can be covered by 'two essentially different principles' shows the '*fictitious* character of fundamental principles' (*Ideas and Opinions*, p. 273, my emphasis) and the 'freely constructive element in the formulation of concepts' (letter to Besso, 6 January 1948, quoted from Holton, p. 231). This, Einstein thought (again, letter to Besso), was not realized by Mach. However we have seen (text to footnote 21 above) how the scientist, according to Mach, 'adds [things] of his own, using his own reservoir of ideas' and so 'constructs' (E 316) 'ideal cases' (E 190f.) or, as Mach also calls them, '*fictions*' (E 418: 'the physicist knows that his fictions present the facts only approximately and using arbitrary simplifications' such as ideal gases, perfect fluids, perfectly elastic bodes and so on). We now notice the very same awareness of a plurality of principles that prompted Einstein to criticize Mach: Einstein did not know, or had forgotten, the complexity of Mach's thought. Cf. footnote 42.

25. *Analyse der Empfindungen*, Jena 1922, p. 18.

26. *Populärwissenschaftliche Vorlesungen*, Leipzig 1896, p. 228. Cf. E 14, footnote: '*psychological* observation is as important a source of knowledge as *physical* observation'.

science. The idea that there exists a sharp boundary between subject and object, mind and matter, body and soul, and the corresponding idea of a real external world untouched by anything 'mental', are ingredients of this kind. In Mach's time these ideas were regarded as unshakeable presuppositions of research (the attitude still survives in an inarticulate form). Mach did not agree: what affects science or is part of science must also be examined by it. Examining the idea of a real external world means either looking for chinks in the mind – matter boundary, or introducing 'contrary idealizations' (p. 263) which are no longer tied to the idea. Mach used both methods.

The second reason for using mixed research strategies was that they led to partial success. Parts of the boundary line turned out to be caused by psycho-physiological processes (Mach Bands), other parts were accidental left-overs from earlier points of view. To examine the matter further and to prepare a science no longer dependent on those accidents Mach introduced his 'monism'. This monism was not part of Mach's *general* views on research, it was a *particular theory* found in accordance with those views *and subjected to them. It was therefore not a necessary boundary condition of research* as is assumed by almost all of Mach's critics, Einstein included. Mach took special care to emphasize this point: 'What is the simplest and most natural starting point for the psychologist need not be the best and most natural starting point for the physicist or the chemist, who face entirely *different* problems or to whom the same problems offer very *different* aspects' (E 12, footnote 1). It is especially misleading to regard Mach's monism as the result of a simple-minded identification of what exists with entities (sensations) which are (a) subjective, (b) basic and (c) not capable of further analysis. Such entities exist neither in science (which has no 'unshakeable' ingredients: E 15, and is 'in need of continuous examination': n. 231, and cf. also E 15 on the difference between philosophical and scientific ways of thinking) nor, therefore, in Mach's monism, which is a scientific theory, not a philosophical principle, as we have seen.

According to Mach, then, the world consists of *elements* which can be classified and related to each other in many different ways (E 7ff.). Elements are sensations 'but only insofar' as we consider their dependence on a particular complex of elements, the human body; 'they are at the same time physical objects, namely insofar as we consider other fundamental dependencies'[27] – 'the elements, therefore, are both *physical* and *psychological* facts' (E 136). They depend on each other in many different ways and there exists no complex of elements that remains unaffected by what goes on outside of it: 'strictly speaking there is no *isolated* thing' (E 15). However, following methods familiar from the sciences, we introduce idealizations or 'fictions' (E 15) such as 'thing' or 'subject' and formulate 'principles' in their terms. The elements are not final – 'they are just as *tentative and preliminary* as were the elements of alchemy, and as are the elements of chemistry today' (E 12). Nor is it necessary to refer every piece of research back to them (E 12, footnote 1: quoted in the preceding paragraph). Formally Mach's monism and the atomic hypothesis have much in common. Both assume that the world consists of certain basic entities, both use scientific research to discover their true nature, both agree that the assumption is not necessary and has to be tested by experience. And just as the atomic theory need not oppose the construction of phenomenological theories, provided the aim is eventually to analyze them in terms of atoms, in the very same manner Mach did not oppose or criticize the gradual building up of a mechanical science provided *its* concepts were not assumed to be final and regarded as the foundation of everything else (p. 483). The difference between Mach and the atomists lies in the basic entities; and here, Mach claimed an advantage. For while it is not possible, according to Mach, to 'explain sensations by motions of atoms' (p. 483), it must be possible to explain atoms in terms of the elements of a perceptual field, or else the atomic hypothesis ceases to be part of an empirical science. Elements as envisaged by Mach are therefore more fundamental than atoms.

5 Einstein's Irrational Positivism and Mach's Dialectical Rationalism

Now compare this sophisticated account of the development and of the elements of our knowledge with a reconsideration of Einstein's description (quoted above in chapter 4, section 3).
        According to Einstein (R 291),[28]

the first step in the setting up of a 'real external world' is the formation of the concept of bodily objects and of bodily objects of various kinds. Out of the multitude of our sense experiences we take, mentally and arbitrarily, certain repeatedly occurring complexes of sense impressions . . . and we correlate to them a concept – the concept of the

27. *Analyse der Empfindungen*, p. 13.

bodily object. Considered logically this concept is not identical with
the totality of sense impressions referred to; but it is a free creation of
the human (or animal) mind. On the other hand, this concept owes its
meaning and its justification exclusively to the totality of the sense
impressions which we associate with it.

The second step is to be found in the fact that, in our thinking
(which determines our expectations) we attribute to this concept of
the bodily object a significance, which is to a high degree independent
of the sense impressions which originally give rise to it. This is what
we mean when we attribute to the bodily object 'a real existence'. The
justification of such a setting rests exclusively on the fact that, by
means of such concepts and mental relations between them, we are
able to orient ourselves in the labyrinth of sense impressions. These
notions and relations, although free mental creations, appear to us as
stronger and more unalterable than the individual sense experience
itself, the character of which as anything other than the result of an
illusion or hallucination is never completely guaranteed. On the other
hand, these concepts and relations, and indeed the postulation of real
objects and, generally speaking, of the existence of 'the real world',
have justification only insofar as they are connected with sense impressions between which they form a mental connection.

What strikes the reader familiar with Mach's writings and with
the history of positivism up to and including the Vienna Circle is
how much closer this account is to positivism than it is to Mach. It
is also much more simple-minded than Mach and most importantly, it is *utterly unreal*. There exists no stage in history, or in
the growth of the individual, that corresponds to the 'first stage';
there is no stage when, surrounded by a 'labyrinth of sense
impressions,'[29] we 'mentally and arbitrarily' select special
bundles of experience, 'freely create' concepts, and correlate the
concepts with the bundles. 'Not only mankind, but the individual, too, finds . . . a complete world view to whose construction he has made no conscious contribution. *Here everyone must
begin* . . . ' (E 5, my italics). 'Commonsense . . . apprehends

28. Einstein's papers quoted in the text here are: 'On the Method of Theoretical Physics', Herbert Spencer Lecture Oxford 1933; 'Physics and Reality',
Journal of the Franklin Institute (1936); 'The Fundaments of Theoretical Physics',
Science (1940). The papers will be quoted from *Ideas and Opinions* giving the
page number of that book preceded by the letter R.

29. When speaking of sense impressions Einstein always means *immediate*
sense impressions. This becomes clear from his letter to Maurice Solovine of 7
May, 1952 (reproduced in A. P. French, ed., *Einstein: A Centenary Volume*,
Cambridge 1979, p. 270, which contains a fascimile of the epistemological part of
the letter). The diagram which accompanies it has at its basis the '*Mannigfaltig-
keit der unmittelbaren (Sinnes) Erlebnisse*' (my emphasis).

the bodies in our surroundings as a whole, without separating
. . . the contribution of the individual senses' (E 12, footnote 1).

However, the 'first stage' not only *does not exist*, it *cannot exist*
as a starting point of knowledge. Mach explains the reasons why:
'Experience alone without accompanying thought would always
be alien to us' (p. 465): a person facing sense experience without
thought is disoriented and unable to carry out the simplest task.
Also, 'an individual sensation is neither conscious nor unconscious. It becomes conscious only by becoming part of the experience of the present' (E 44): the articulation of these experiences
is a presupposion of their being conscious and therefore cannot
be achieved by an ordering process applied to the conscious but
not yet articulated field of sensations. 'Imagination gets hold
already of a single observation, changing it and adding to it' (E
105); and this is necessary, for '*understanding* nature must be
preceded by *grasping it* in the imagination so that our concepts
may have a living and intuitive content' (E 107): concepts, too,
cannot be 'pure', they must be suffused with percepts before they
can be used for ordering anything. Neither concepts nor sensations can first exist separately, then combine and in combining,
form knowledge.

Further, there is no sharp boundary between memory and
imagination – no experience is so isolated that other experiences
cannot influence the memory of it. Memories, however, are
'poetry and truth combined' (E 153: the reference is to Goethe's
autobiography *Dichtung und Wahrheit*): 'observation and
theory, too, cannot be sharply separated' (E 165), and so 'the
two processes, *the adaptation of our ideas to facts and the adaptation of our ideas to each other* cannot be sharply separated.
Already the very first sensations [of an organism] are co-deter-
mined by the innate and the temporary state (*Stimmung*) of the
organism [which depends on biological needs as well as on tradition: E 70ff; cf. E 60] and later impressions are influenced by
what went on before' (E 164). The whole complex of perceptions
that arises in this manner 'is organically older and better founded
than conceptual thought' (E 151); and common sense which, 'to
start with, cannot at all be separated from scientific ideas' (E 232)
not only knows no sensations in the sense of Einstein, it could not
possibly have formed such a complex and abstract idea (E 44,
footnote 1).

Both Mach and Einstein believed in a close connection
between science and common sense. 'Scientific conceptions are

immediately connected with common sense ideas from which they cannot at all be separated' (E 232). This is the reason, says Einstein (R 290), why scientists 'cannot proceed without considering critically a much more difficult problem [more difficult than the analysis of scientific ideas, that is], the problem of analysing the nature of everyday thinking and of changing it where a change seems necessary.' But the way in which Einstein describes the situation makes the changes seem rather easy and suggests the wrong methods for carrying them out. If 'sense experiences are the given subject matter' (R 325),[30] if the concepts used for bringing order into this subject matter are 'arbitrary', 'free creations' and 'essentially fictitious' (R 273), then all we need to do is to abandon one set of fictions, 'freely invent' another and a third, and a fourth, compare how well they all order the sensations and choose the set that does the best job. The procedure may be long and tedious, but it has no intrinsic difficulty. All it involves is a 'free play with concepts'.[31] If, on the other hand, there is no neat separation between sensations, imagination, thought, memory, fantasy, genetics, instincts (E 164; cf. E 323: 'scientific hypothesizing is only a further development of instinctual primitive thought'), a dream and waking (E 117); if any historically given subject matter is an alloy of all these entities or perhaps no alloy at all, but a simple unitary thing[32] (which would mean that sensations are not subject matter, but 'fictions') – then research will be very different indeed from the procedure implied in Einstein's description. The invention of new principles will not be as 'free' as assumed by Einstein nor will it suffice simply to reshuffle some familiar pieces, for the existence of the pieces themselves will now be in question.

It is strange to see how inventive scientists such as Einstein and

30. Remember that in his letter to Besso (footnote 2 above) Einstein ascribed this very assumption to Mach (who never held it) and criticized it. In the present quotation, 12 years before the letter, he accepts it himself – and so he does in his autobiography (Schilpp, op. cit., 6ff.) which was written in 1946 and in the letter to Solovine, written 1952, cf. footnote 29: sensationalism has a way of insinuating itself when sensations are assumed to be incapable of analysis.

31. Schilpp, op. cit., p. 6.

32. Mach, for example, seems inclined to count some of Duhem's 'qualities' among the elements. Cf. his introduction to the German translation of Duhem's main work on the philosophy of science, *Ziel und Struktur der Physikalischen Theorien* (Leipzig 1908). But 'qualities', for Duhem, are things like currents, charges and so on.

Planck,[33] who vigorously opposed positivism, still used an essential part of it and so gave an account of science that was much more simple-minded than the practice in which they were engaged. Ernst Mach, the alleged positivist, was one of the few thinkers to recognize the fictitious character of the description and to replace it by a more realistic account. In this he became one of the forerunners of gestalt psychology, constructivism in mathematics, Piaget, Lorenz, Polanyi and Wittgenstein (who, unfortunately, is much more longwinded than Mach) as well as of more recent attempts to look for patterns in the physical world. physical world.

Planck and Einstein not only used parts of positivism which Mach had criticized, they not only opposed Mach in areas where no real conflict existed, they occasionally even opposed each other or at least made statements implying such an opposition. Thus Planck in stating the realist doctrine added 'here we cross out the positivistic "as if" ' (Planck, p. 234) which had been used by Einstein in his *criticism* of the positivistic doctrine; while Einstein when emphasizing the 'real existence' of bodily objects (R 291) meant concepts 'which [are] to a high degree independent of sense impressions' — which agrees with Mach (see Section 4 above)[34] but not with Planck for whom reality was an ontological and not a semantic problem. Planck's 'faith' in an external world seems to contradict Mach, but Planck added that instead of speaking of a 'faith' one might also speak of a 'working hypothesis' (Planck, p. 247) – a legitimate procedure in Mach's philosophy (E 143).[35] Both Planck and Einstein used formulations that seemed to put them in opposition to Mach but are found to agree with him when attention is paid to the different ways in which key terms are used.[36] Altogether the battle about Mach and positivism was a net of confusions, with the enemies adopting part of the very doctrine Mach was supposed to have held but did not hold and calling each other positivists when in

33. For Planck, cf. his *Vorträge und Erinnerungen*, Darmstadt 1969, p. 45 (sensations are the 'recognized source of all our experiences'), pp. 207, 230 (all concepts of physics are taken from the sensory world and are improved and simplified by relating back to it), p. 226 (contact with the sensory world must always be maintained), p. 229 ('the source of all knowledge and the origin of every science lies in personal experiences. They are immediately given, they are the most real thing one can conjure up and the first starting points for the trains of thought that constitute science'), and p. 327 (the real world is only perceived through the medium of the senses).

34. Note, however, that Mach in his discussions starts from individual *physical* events, not from sensations.

fact they were trying to get away from it. Mach and Planck were even one in *asserting* that talk of the 'fictitious and freely invented character' of theoretical concepts prevented a true understanding of the role of fundamental principles – but they had very different reasons. For Planck fundamental principles were neither fictitious nor arbitrary *because they described real features of the real external world*. For Mach they were neither fictitious nor arbitrary *because they depended on many historical factors, instinct among them*. Planck and Einstein, of course, also recognized the need for an agency that moves research from one stage to another and they called this agency intuition (R 226) or faith (Planck, passim, and cf. footnote 35). However there is a great difference between intuition, or faith as conceived by Planck and Einstein, and Mach's instinct.

According to Mach, 'the true relation between various prin-

35. There is an interesting difference between the various statements on realism Planck made in the course of his life. In his paper of 1908 which led to the famous exchange with Mach he calls realism a 'faith, as hard as a rock' (p. 50), and held by the majority of physicists. He also objects to regarding the subject-object distinction as a practical (conventional) boundary (p. 47), implying that it expresses a bifurcation of reality itself. In 1913 he again mentions the 'faith' but adds: 'However faith alone does not suffice for as the history of science teaches us, it can lead us astray and lead into narrowmindedness and fanaticism. To remain a trustworthy leader it must be constantly checked by logic and experience' (p. 78). In 1930 he again calls realism a faith but adds that 'speaking more carefully, one might also call it a working hypothesis' (p. 247). But in the very same paper he 'cross[es] out the positivistic "as if" and ascribe[s] to the so-called practical inventions a higher degree of reality than to direct descriptions of the immediate impressions of the senses' (p. 234). He also emphasizes the 'metaphysical' (p. 235) or 'irrational' (p. 234) elements introduced by a realist faith. In 1937, in his address 'Religion and the Natural Sciences' he sets out to examine 'what laws are taught to us by science and what truths are untouchable' (p. 325) and he mentions the existence of a real world independent of our impressions (p. 327) as well as the existence of laws in this real world (p. 330). More cautiously he mentions the principle of least action and the existence of constants among these laws.

It seems that Planck wavered between a more scientific (i.e. testable) version of realism – a version, incidentally, which Mach himself had proposed (*Mechanik* p. 231) — and a more 'irrational' or 'metaphysical' version. This wavering may have been induced by his twin loyalty to science which for him always meant testability (was this a remnant of Mach's influence?) and also to a certain philosophical faith, and it may have been encouraged by his recognition that while one may postulate a reality, even an 'absolute' reality, one must at the same time admit that no particular scientific result or principle can ever completely capture it (p. 182 – with a fine reference to G. E. Lessing).

36. For example, Einstein argues against 'abstraction' as a principle of science (R 273), pointing to the creative aspect of theory building, while according to Mach abstraction is 'a bold intellectual move' (E 140).

ciples is historical' (p. 73). 'The strictest and the most complete account of an idea consists [therefore] in clearly exhibiting all the motives and all the paths that have led up to it and have confirmed it. The *logical* connection of the view with older, more customary and *uncontested* ideas is but part of this procedure' (E 223). 'Getting in touch with the classical authors of the renaissance of scientific research provides incomparable pleasure and thorough, lasting, irreplaceable instruction for precisely the reason that these great and naive human beings in the charming joy of their inquiries and their discoveries tell us, without any learned mystification (*ohne jede zunftmässige gelehrte Geheimnistuerei*) what has become clear to them and how it became clear to them. Thus, reading Copernicus, Stevin, Galileo, Gilbert, Kepler we learn without grandiloquence (*ohne allen Pomp*) the guiding motives of research explained by examples of the greatest successes in research . . . A cosmopolitan openness . . . characterizes the science of this time' (E 223ff.).

It is well known that modern philosophers of science have chosen to disregard this cosmopolitan openness and have replaced it with what they call, somewhat optimistically, a 'rational' account, or a 'rational' reconstruction. A rational account, to use Mach's words (last paragraph), explains an idea by showing 'its logical connection with older, uncontested views', without indicating how these views arose and why they should be accepted. Now Einstein's and Planck's accounts of knowledge were rational accounts in this sense. They accepted a set of 'uncontested' entities (sense data) and a set of uncontested views ('logic') and they made assertions about the relation of real scientific theories to these unexplained and uncontested things. Naturally, they left many questions unanswered.

Moreover, they used fictions, not real things, to answer the questions they *did* consider: there are no immediate sensations anywhere in science or in commonsense (sensations do occur in psychology, but as theoretical entities, not as a subject matter shared by all sciences). The 'arbitrary and fictitious character' of the principles is but a reflection of this unreal and fictitious starting point: existing principles of science, compared in an incomplete fashion with fictitious entities supposed to be their one and only subject matter, will naturally appear arbitrary and fictitious themselves. And an agency that connects the fictions with reality – Einstein's 'intuition', Planck's 'faith', Dirac's 'beauty'[37] – is bound to have some very strange properties. This is why Planck emphasized the 'irrational' and 'metaphysical' character of the basic principles and why Einstein spoke of 'the religious basis of scientific effort'.[38] *But this irrationality, this intrusion of religion into an allegedly rational enterprise such as science, is but the mirror image of the lingering positivism of Planck and Einstein*, it is a reflection of the incompleteness of their views of knowledge. It does not arise in Mach.

For although Mach ascribes to instinct a very important function, and although he emphasizes its 'power' and 'higher authority' (p. 26) and the fact that it is 'something alien and free of subjective elements' (p. 73), although he shows how progress is made by reforming facts on the basis of instinct and declares the greatest scientists to be those who 'combine the strongest instinct with the greatest conceptual power' (p. 27), there are two essential differences between him and Planck-Einstein. First, instinct does not operate on sensations, it operates in a concrete historical situation that is only partly articulated in the form of postulates, standards and experimental results but largely consists of unconscious tendencies (of thinking, perceiving, reacting to reports). It is a real agency working in the real world. Secondly, 'instead of engaging in mysticism', Mach asks 'how do these instinctive pieces of knowledge arise . . . what is contained in them?' (p. 27) and 'what is the source of their greater authority' (p. 26) — greater, that is, than experimental results? The answer to the last question is simple: an instinct that propels the sciences *presents itself* as being independent of our actions and beliefs while every experiment we perform depends on assumptions we have formed ourselves and which therefore *are known* to express our (subjective) expectations. The question of the trustworthiness of the authority is answered by reflecting on its *form*: instinctive knowledge 'is mainly negative, it does not tell us what must occur, it rather tells us what cannot occur' (pp. 27ff.).[39] Its *content* becomes clear from the fact that the forbidden events 'strongly contradict the unclear mass of experiences in which individual events cannot be distinguished' (p. 28) – they contradict the *expectations* of a person or a group of persons in a certain stage of adaptation to the world. The content has *authority* because instinctive knowledge, though rather vague when compared with detailed experimental results, 'rests on a broad foundation' (E 93). We know from innumerable

37. Cf. for example 'The early years of relativity' in Holton and Elkana, op. cit., p. 83.

38. In his 1929 essay 'Über den gegenwärtigen Stand der Feldtheorie', quoted from Holton, p. 242.

experiences, only some of them consciously formulated, that heavy objects do not rise on their own account however complex the physical connections made between them, that objects of equal temperature when brought in contact retain this temperature, that differences of temperature, pressure, density equalize out but not the other way around – and so on. We now also realise that the authority of instinctive principles increases whenever contact with the material world increases which is a reason why the sciences are bound to profit from the (explicit and instinctive) knowledge of artisans (E 85). It is to be admitted that 'a carefully conducted experiment provides many details' but the 'most secure support' for the sciences 'arises from being related to the crude experiences' contained in instinctive principles. 'This is how Stevin adapted his quantitative ideas about inclined planes and how Galileo adapted the quantitative ideas about free fall to their general experience in the form of thought experiments' (E 193), this is the reason why 'special quantitative points of view should be tentatively adapted to general instinctive impressions' (p. 29) and this is also the rationale – given by Mach, not by Einstein – for Einstein's refusal to be intimidated by any 'verification of little effects'[40]: *instinctive knowledge, having passed tests by a great variety of qualitatively different experiences overrules a novel experiment that is based on special assumptions in a narrow domain only.*[41] Again we notice an agreement on procedure, but the reasons given by Mach and by Einstein are very different indeed.

As we have seen, Einstein emphasized the arbitrary and fictitious character of general principles. What he meant was that there is no logical way from experience (and that, for him, meant immediate sensations) to principles (R 273). With this narrow interpretation Mach might have agreed, for he not only noticed but also emphasized the (logical) conflict between principles and special experiments and he advised scientists to adapt the latter to the former and not the other way around. But Mach did not admit that the principles were therefore 'freely created by the human mind', and rightly so, for there are many constraints beyond those allegedly imposed by logic, and 'rational' action not merely attends to these additional constraints but is also

---

39. Instinctive knowledge shares this form with the laws of nature. They are 'restrictions which, being guided by experience, we impose on our expectations' (E 44ff. – original all italics); 'the progress of science leads to an increasing restriction of our expectations' (E 452).

carried along by them.

Now when Mach argued against 'free creations' he gave a description and issued a warning. The description mentions what has just been said. 'One often calls the numbers "free creations of the human mind". The admiration for the human spirit that is expressed in these words is quite natural when we view the finished and imposing edifice of artihmetic. However, our understanding of these creations is better served by tracing their *instinctive beginnings* and considering the cirumstances which led to the need for these creations. Perhaps one will then realise that the first structures (*Bildungen*) which arose here were unconsciously and biologically *forced* upon humans by material circumstances and that their value could be recognized only after they had come into existence and had proved useful' (E 327). Talking of 'free creations' (or 'bold hypotheses') disregards this complex net of determinants, replaces it by a naive and fictitious account and deceives the researcher about his task. For – and this is Mach's warning – any train of thought that leads away from instinct loses touch with reality and causes 'dreamlike excesses and unfortunate monstrous special theories' (pp. 29f.).

6   Atoms and Relativity

There is no doubt that Mach counted atoms among the monsters that had arisen because scientists had wandered away from the instinctive basis of the science of their time. Atoms, for Mach, were not merely idealizations, but 'pure objects of thought

40. Letter to Besso of March 1914 quoted from Carl Seelig, op. cit., p. 195. Cf. also the *Born – Einstein Letters*, p. 192: 'It is really strange that human beings are normally deaf to the strongest arguments while they are always inclined to overestimate measuring accuracies.' 'Human beings', of course, means 'physicists'.

41. A critical experimental investigation is not excluded thereby: quite the contrary, Mach demands it (p. 231) – but it must provide a basis comparable to the basis implicit in the principle.

According to Mach theories confirmed in a wider domain also overrule *theories* confirmed in a more narrow domain: 'If some physical facts should require a modification of our concepts then the physicist will prefer to abandon (*opfern*) the less perfect concepts of physics instead of the simpler, more perfect, stronger concepts of geometry which are the most solid foundations of all his ideas' (E 418). Note that confirmation, for Mach, always includes the possibility of failure and always includes instinct and unconscious trials and comparisons. Note also how this approach can be used to defend an empirical medicine against encroachment by theories which rest on a few sophisticated, though remote experiments.

(*blosse Gedankendinge*) which from their very nature *cannot impinge upon the senses* (*nicht in die Sinne fallen können*: E 418). An idealization such as an ideal gas, a perfect fluid, a perfectly elastic body, a sphere (E 418) can be connected with experience by a series of approximations – it obeys a *principle of continuity* (p. 131). Even a complex and comprehensive theory such as Newton's mechanics contains statements which can be gradually transformed into descriptions of observable matters.[42] Indeed, Kant's entire phenomenal world with its rich texture obeys a principle of continuity as understood by Mach. Pre-Kantian substances, on the other hand, or Kant's *Ding an sich* don't have this property (p. 466). They can be neither experienced, nor linked to experience by a chain of approximations, idealizations, abstractions; they are pure constructions of thought (*reine Gedankendinge*) and statements about them are *untestable in principle*. Mach assumed (assumption A) that entities of this kind have no place in science.

He also assumed (assumption B) that atoms as conceived by the majority of the atomists of his time had this undesirable property. His objection to atoms was based on these two assumptions, not on a naive positivism as almost all his critics, Einstein among them, assert.[43]

Assumption A is accepted by Einstein (see the text to footnote 11 and related passages), by most scientists and by almost all modern philosophers of science. It is the only methodological (epistemological) assumption in Mach's argument. It is therefore not possible to criticize Mach's rejection of atoms on methodological grounds.

Assumption B is a historical assumption and like all historical assumptions it is not easy to establish. Besides, Mach admits that 'the atomic theories might enable scientists to represent various facts' but urges them to regard them as '*provisional aids*' and to strive 'for a more natural point of view' (p. 466). Now this is precisely what Einstein did in his early papers on statistical

42. Einstein's comment, in the letter to Besso quoted in footnotes 2 and 24 above, that Mach 'did not know that this speculative character belongs also to Newton's mechanics' (Holton, p. 232) shows that he had not read his Mach very well or had long since forgotten what he had read.

43. Mach was of course aware of the existence of calculations that connected mechanics with thermodynmical laws and facts. However he suspected that many of these calculations were *ad hoc* ('specially invented for the purpose' – p. 466) and that genuine explanations could be achieved without using the details of the mechanical point of view. He was right on both counts – cf. the text to the next footnote.

phenomena. In these papers he not only criticized the existing
kinetic theory because it had 'not been able to provide an
adequate foundation for the general theory of heat',[44] he also
tried to free the discussion of statistical (heat) phenomena from
special mechanical assumptions and he showed that some very
general properties suffice to obtain the desired results.[45] Furthermore, Einstein realized the need for stronger arguments[46] and so
constructed, in his paper on Brownian motion, the very same
kind of continuous connection between atoms and experience
which Mach had demanded, and thereby removed a major flaw
of the existing atomic theories.[47] Finally, the quantum theory
introduced the 'more natural point of view' Mach had been
looking for.

We have to conclude that Mach's criticism was reasonable,
that it had nothing to do with a positivistic attitude, that Einstein
(and von Smoluchowsky and the founders of the quantum
theory) acted as if it had been reasonable, improving the kinetic
theory in precisely the respects Mach had criticized, and that
Einstein's later criticism of Mach on atoms cannot be taken
seriously. If there was any positivism anywhere it occurred in
Einstein, not in Mach.

The special theory of relativity satisfied the principle of continuity as defended by Mach. We have seen that and why the
popular explanations of Mach's objections to this theory cannot
be accepted: they criticize Mach for views he never held and by
contrast praise Einstein for having used procedures explicitly
recommended by Mach. Yet here, too, the solution seems to be

44. 'Kinetische Theorie des Wärmegleichgewichtes und des zweiten Haupt-
satzes', *Ann. Phys.*, 9 (1902), 417.

45. The properties are: first order linear differential equations for the
temporal variation of the state variables (Mach, following Petzold, had regarded
this as an important *empirical fact*: cf. footnote 19), a unique integral of motion
(the energy) and an analogue to Liouville's theorem (which Mach might have
regarded as just 'another side of the same [basic mechanical] fact': 454). Cf.
Martin Klein, 'Fluctuations and Statistical Physics in Einstein's Early Work', in
Holton and Elkana, op. cit., esp. p. 41.

46. Schilpp, p. 46.

47. Loc. cit. It is reported that (Einstein, op. cit., p. 48 to the contrary)
Brownian motion did not convert Mach. However we do not know whether Mach
read Einstein's argument as an argument from continuity or whether he was just
informed of the identity of constants obtained from Brownian motion on the one
side and Planck's formulas on the other (Einstein, loc. cit., suggests the latter).
Only the first argument would have convinced him – and rightly so, for an
agreement of *numbers* does not tell us anything about the *entities* to which the
numbers belong.

fairly simple and Mach himself has indicated where it can be found. In his short comments in the introduction to the *Optik*[48] Mach gives three reasons for his criticism: the increasing dogmatism of the defenders of relativity, 'considerations based on the physiology of the senses' and 'conceptions resulting from . . . experiments'.[49]

The charge of dogmatism was well founded. It applied to Planck who, despite occasional cautionary clauses in his writings (cf. the quotations in footnote 35), seems to have regarded relativistic invariants as parts of the absolute reality he postulated behind the world of science and the sensory world. It applied to most other physicists who, unaware of Mach's examination of the subject/object boundary and unwilling to engage in a critical examination of so fundamental a prejudice, practiced 'physics as usual'; which means, they took the boundary for granted and used relativity for stabilizing it and making it more precise. The 'considerations' are most likely connected with Mach's interdisciplinary examination of the boundary (cf. text to footnote 26 above), his conjecture that 'mental' events have 'material' ingredients (and vice versa), and his attempt to develop a point of view that took these matters into account. The conflict between Mach and the uncritical followers of Einstein was, therefore, first, a conflict between two different scientific theories, one containing a well defined subject/object boundary, the other redefining or dissolving this boundary in accordance with the results of scientific research in physics, physiology, psychology. It was, secondly, a conflict of attitudes: Mach wanted to *examine* the matter, his opponents either *regarded it as settled* or did not even realize that there was a problem. Again Mach emerges as a critical scientist trying to make some very fundamental parts of science conform to principles already practiced in other regions of scientific research. It is also well known that the conflict has continued right into the opposition between Einstein and Bohr on the quantum theory and the more recent clash between space – time theories and quantum theories.[50]

About the 'experiments' which Mach mentions at the end of his critical comments (cf. the brief excepts from the *Optik* in the text above), we know nothing.

---

48. But see the *Afterword* to the present essay for a comment on the authenticity of this text.

49. *The Principles of Physical Optics* (Dover publications).

7   Lessons to be Learned

What do we learn from this sketch of the Mach – Einstein episode? We learn, first, that one cannot trust received opinions, or received versions of 'great turning points of science' or 'great debates', even if they should happen to be supported by out-standing scholars in the relevant fields. We learn, secondly, that the faults of the received opinions can often be found without detailed archival studies – a careful reading of a few well known books suffices. Such reading makes us realize, thirdly, that the received versions most of the time are not only incorrect, but much simpler (and much more simpleminded) than the events they describe. Thus we are led to suspect, fourthly, that many so-called 'great issues', such as the issue between 'realism' and 'positivism' that is usually connected with the Mach – Einstein conundrum, are sham battles caused by misunderstandings and oversights, and that, taken by themselves and without a historical analysis (of the simple kind mentioned under point two), they teach us absolutely nothing about science and knowledge in general. From this we learn, fifth, that philosophical systems claiming to have solved such issues are indistinguishable from the productions of con-artists except for the fact that con-artists know what they are doing while the philosophers do not. And all the philosophical battles that arise when different systems, trying to give different accounts of the same 'great issue' or of the same 'great step forward' or of the same 'revolution', confront each other with reasons and counter-reasons are ever so much moon-shine.[51] The moonshine, of course – and this is the sixth point – is a boon to all those who, lacking the talent to understand and influence a complex historical process, can now be simple-minded and still remain philosophers and can even claim to be more 'rational' than those who are dissatisfied with their naive models. This encourages us, seventh, occasionally to go a little further, to take a critical look at the events allegedly constituting the issues and to 'rescue'[52] the participants from the fairy tales that are being told about them. Such an activity is very inter-

50. This conflict, too, is a conflict between physical assumptions overlaid by a critical attitude on one side (Bohr) and dogmatic obstinacy on the other (Einstein). Cf. my essay on Bohr in Vol. i of my *Philosophical Papers*, Cambridge, 1981.

51. For the moonshine created in the case of Bohr, see my essay on Bohr referred to in footnote 50. Part of the moonshine surrounding the so-called Copernican Revolution is discussed in my *Against Method* (London 1975), third largely rewritten edition *Wider den Methodenzwang* (Frankfurt 1986).

esting in itself, for it creates great and often quite unsuspected surprises. It is necessary, if our history and our philosophy are to be more than daydreams posing as the outlines of a real world.

Afterword, 1985

It now appears that the foreword to the *Physikalische Optik* and the foreword to the 9th edition of the *Mechanik*, which contain passages critical of the special theory of relativity, were written by Ludwig Mach, Ernst Mach's son, and inserted without Ernst Mach's knowledge. In a word, both texts are a fake. The evidence, which is strong though circumstantial and which to me seems entirely convincing, has been assembled by Dr. Gereon Wolters of the University of Konstanz. I accept his conclusions and the interpretation he bases on them: see 'Atome und Relativität – Was meinte Mach?', in R. Haller and E. Stadler, eds., *Ernst Mach: Leben, Werk und Wirkung*, Vienna 1986. My remarks on Mach and atomism remained untouched by these discoveries.

52. The great German poet and philosopher G. E. Lessing wrote a series of '*Rettungen*' trying to rescue great and much maligned historical figures from the injustice done to them by clergymen, scholars and general rumour.

8

Some Observations on Aristotle's Theory of Mathematics and of the Continuum

1.   In book ii/2 of his *Physics* and in book XIII/3 of his *Metaphysics*, Aristotle explains the nature of mathematical objects.[1] The explanation is fairly simple, but long and elaborate discussions are added to combat alternative views and to eliminate mistakes. I will not go into these discussions, nor will I mention and comment upon the modern debates about their correct interpretation. I will merely present Aristotle's statements, add clarifying remarks, examine consequences in physics, compare them with objections by later authors, and show how they are related to modern problems. In quoting Aristotle I will omit special reference to the *Physics* and to the *Metaphysics* – here the page numbers speak for themselves. Simple numbers in parentheses such as (14) refer to the sections of the present paper.

2.   Physical bodies, says Aristotle, have surfaces, volumes, lines, and points. Surfaces, volumes, lines, and points become the subject matter of mathematics *by being separated from bodies* (193b34).

We also read (1964a28ff.) that physics 'deals with things that have principles of movement in themselves; mathematics is theoretical and is a science that deals with things that last – *but are not separate.*'

'They *cannot in any way exist separately* – but they cannot exist *in* sensible objects either' (1077b15ff.; cf. 1085b35ff.).

The contradiction is resolved by realizing that 'things are said to be in many different ways' (cf. *Met*.iii/2 and numerous other

1. An early version of the paper was read by Dr. Rafael Ferber who sent an extended commentary and suggestions for improvement. I have adopted some of his suggestions and rewritten the text accordingly.

passages). Mathematical objects have separate existence in some of these senses but not in others.

Assume that to exist means to be an individual entity that does not depend on other objects and is as real as (or perhaps even more real than [1028b18]) physical bodies. If mathematical objects exist in this sense, they cannot be *in* physical objects – for then we would have two objects in one place (1076a40ff.; cf.998a13f.) – nor can they be physical objects ('sensible lines are not lines like those the geometer describes: there is nothing perceptible that is straight or curved in [the strict geometrical] sense, the circle touches the ruler not at a point but after the manner of Protagoras' (998a1ff.). Neither can we assume that physical objects are *combinations* of mathematical objects: a combination of static and unperceived objects only yields further static and unperceived objects (1077a34ff.) and joining physical material (bronze, for example) and mathematical shape (an individual sphere, for example), both conceived as complete and self-sufficient individuals, yields a pair of complete and self-sufficient individuals (bronze; sphere), not a single individual having sphericity as a (dependent) property (1033b20ff.). Aristotle gives further arguments, not all of them admirable, to show that mathematical entities, interpreted as complete and independent individuals (or as 'substances', in Aristotle's terminology), are neither *in* physical objects nor *outside* of them and separate from them.

3. Although it is not possible to have self-sufficient *objects*, either in physical bodies or apart from them, that lack important properties of physical bodies, it is possible to have incomplete *descriptions* of such bodies.

> For example . . . . there are many statements about objects considered merely insofar as they can be in motion which say nothing about their essential and their accidental properties without it necessarily following that there is motion apart from sensible things or that there is a distinct moving entity in them (1077a34ff.).

Similarly, 'there are properties and domains of knowledge that treat [changing objects] not insofar as they are moving but insofar as they are bodies or only insofar as they are planes and lines and are divisible, or as indivisibles having position, or as indivisibles only' (1077b23ff.). 'The same principle applies to harmonics and optics for neither of these sciences studies objects as sight or as sound but as lines and as numbers – but the latter are attributes belonging to the former. And the same is true of mechanics' (1078a14ff.). No error arises from this 'any more than when one draws a line on the ground and calls it a foot long when it is not – for the error is not included in the premises' (1078a18f.; cf. Berkeley's very similar account in the introduction to his *Principles of Human Knowledge*, 1710).

Bearing this in mind, we can say not only without qualification that what is separable exists but also that what is inseparable exists and is separable by description. For example, the objects of geometry exist; *they are sensible objects* but only accidentally so, for the geometer does not treat them as sensible objects (1078a1ff.).

4. Not all incomplete descriptions of bodies disregard motion and perceptibility. 'Straight' and 'plane' do; 'torn' and 'polished' do not – the last two descriptions contain an implicit reference to changeable physical material (Aristotle's favourite example is *simós*, meaning snub-nosed, as opposed to 'concave': 'the definition of *simós* contains the matter of the object, for snubness is only found in noses, whereas the definition of "concave" does not' [1064a23ff.]). As a result, 'straight' and 'plane' can be separated in the sense just explained, while 'torn' and 'polished' cannot. Aristotle criticized Plato for separating things such as 'flesh', 'bone', and 'man' that belong to the latter category (194a6ff.; cf. 1064a27ff.) and for trying to define 'lines [which are separable] from the long and short [which are not], planes from the broad and narrow and solids from the deep and the shallow' (1085a9ff.).

5. If a feature or a property or an entity is separable in the sense explained, then we can either treat it as separated, i.e., we can discuss it without paying attention to other features *that are also present*, or we can give a more complete description. However, we cannot do both at the same time, i.e., we cannot mark or subdivide a physical object by a mathematical point or by a mathematical plane. It is customary in physical calculations to 'imagine' a physical object O 'cut' by a mathematical plane P or to 'consider' a volume V in it (Figure 1). According to Aristotle, this is nonsense. Mathematical planes cannot subdivide physical objects; only physical surfaces can, and those have physical properties not present in the undivided physical body. If P is to divide O, then P must be identifiable as a surface – there must be a narrow slit at *P*: 'whenever bodies are joined or divided, their boundaries instantaneously become one at one time namely, when they touch, and two at another, when they are divided. Thus when the bodies are combined the surface does not exist but has perished' (1002b1ff.). One should compare this account with what the quantum theory says about location and division.

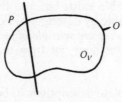

Figure 1

6. These considerations play an important role in Aristotle's theory of place. According to Aristotle, the place of an object is the 'inner limiting surface of the body that contains' the object (212a7), this inner limit being a real physical subdivision. Hence, 'if a thing is not separated from its embracing environment but continuous with it, then it has no place in it, but it has a place as part of the whole' (211a29ff.). For example, a bottle partly filled with water and floating in a lake has a place in the lake: the surface consists of the surface where the outside water meets the bottle – added to the surface where the outside air meets the bottle (Figure 2). The water inside the bottle also has a place, viz., the inner surface of the glass where it touches the inside water plus the surface of the inside air that touches the water. Both places are physically identifiable surfaces, however; a drop of water inside the bottle is *part* of that water; it has *no place* in that water; it has only, as part of that water, a place inside the bottle. We may say that the droplet potentially has a place inside the water of the bottle and that this potential place can be actualized when the droplet is physically separated from the rest of that water, e.g., when it freezes (212b3ff.).

The argument of 3 also shows that place cannot be an inner extension (*diástema*) of a body that stays behind when the body has departed (as place is supposed to stay behind [208b1ff.]), for, if there were such an entity, 'there would be infinitely many places in the same thing' (211b21). This argument, says H.

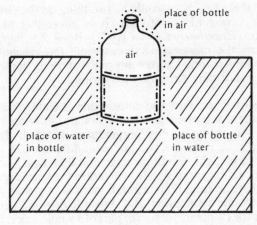

Figure 2

Wagner (*Aristoteles Physikvorlesung*, Darmstadt 1974, pp. 544ff.) 'is one of the great cruces the Aristotelian physics has offered to its interpreters since antiquity.' But the situation is rather simple, almost trivial. In the argument, the *diástema* of a particular body is viewed physically, not mathematically (place, after all, exerts a physical influence [208b11]), and it is supposed to be identical with place. Viewing the *diástema* physically implies that a real physical object is associated with it; identifying the *diástema* with place means demanding that this object behave like a place, i.e., that it stay behind when the body whose place it is supposed to be moves on (208b1ff.). Staying behind the *diástema* does not turn into a mathematical entity, nor does it change from a particular feature of a particular body into something that is shared by all bodies (every physical object has its own individual place and therefore its own individual *diástema*). Hence, every place, having been occupied and left by many different objects, contains many different *diastémata*; and, as every place is occupied by a body (there is no vacuum) and every *diástema* is a place, every body will contain an infinity of places.

7.   Similar remarks apply to the idea that the void might be an independently existing entity, similar to a *diástema* (216a23ff.). Local movement consists in one body replacing other bodies. The void was introduced not in order to be replaced by a body,

but to receive it (213b5f.). It can, therefore, receive a wooden cube. In this case, there would be two things in the same place – which is impossible (216b11f.). It is interesting to see that Guericke (*Experimenta Nova* [1672], Book 2, Chapter 3) re-introduces the *diástema* as representing the vacuum without mentioning Aristotle's arguments to the contrary but with heavy sarcasm concerning Aristotle's philosophy.

8. Aristotle's account also defuses one of Galileo's arguments against the assumption that heavy bodies fall faster than light bodies. According to the argument a heavy body can be imagined as consisting of two bodies of different size, one small and therefore light, the other large and therefore heavier. The light part, falling more slowly, will retard the heavier part which means that the total body will fall more slowly than part of it, contrary to the assumption. But Aristotle does not permit us to consider the (separate) action of part of a whole unless the part is physically separate from the whole. But then the argument no longer works.

9. Aristotle's view of mathematics has especially interesting applications in the domain of motion. In 5 we saw that, in order to make physical sense, statements about parts and subdivisions must be about physically identifiable separations: a continuous body has no actual parts unless it is cut and its continuity inter-rupted thereby. Applied to motion, this means that part of a continuous motion can be separated from another part of the same motion only by a real modification, this time of the motion; i.e., the motion must either slow down or come to a temporary halt; it must 'stop and begin to move again' (262a24f.). This is how Aristotle solves one of Zeno's paradoxes. Zeno had pointed out (263a4ff.) that a movement over a certain distance must first cover half the distance, then half of the half and so on – which means that the motion never gets completed. According to Aris-totle, one subdivides the motion either by using mathematical points – then no subdivision has occurred — or by using physical ('actual' [263b6ff.]) points – then the subdivision changes the motion, turns it into an 'interrupted motion' (263a30f.), which is indeed never completed.

10. Few people are satisfied with this solution. The reason is that the idea of motion usually connected with the paradox differs from the idea Aristotle uses for solving the paradox. Aristotle, the critics feel, does not meet the paradox head-on, he evades it. And he evades it in a particularly simpleminded fashion, introducing acts of subdivision when the problem is the nature of a motion that proceeds without outside interference. The assumption behind the feeling is that the idea of such a motion is without fault, that the paradox arises from a mistaken use of it, and that the task is to correct this mistake and not to start talking about entirely different processes.

If this assumption is incorrect, i.e., if the view of motion involved is inadequate and perhaps even incoherent, then its replacement is not an evasive move but a required one. Zeno's argument will then cease to be a mere paradox and will become a further contribution to its downfall. The main question is: What is the view of motion that those who speak of evasion have in mind, and how can it be defended?

Briefly, the view can be stated as follows: for every point A of the line to be transversed, the event 'passing point A' is part of the motion *whether we now interfere with it or leave it alone.* Motions consist of individual punctiform events of this type, and lines consist of individual points. This is an interesting cosmological hypothesis, but is it acceptable? Aristotle says no, and his main reason (which will be discussed in greater detail in 19 below) is very simple: a continuous entity such as a line or a continuous motion is characterized by the fact that its parts are connected or hang together in a special way. Indivisible entities such as points or passings of points cannot be connected in any way whatsoever; hence, lines cannot consist of points and continuous motions cannot consist of passings of points.

Similar though more complex arguments are provided by the quantum theory that states that we can have pure motion (well-defined momentum) but without any passing of points or a precise passing of points, but then we have no longer any coherent motion.

The assumption is, therefore, incorrect, the view of motion implied impossible, and its removal a necessity, not an evasion.

11.   If motion can only be subdivided by modifying it, then any clearly marked subdivision must be accompanied by a temporary change of motion; for example, the stone thrown upward must come to a halt at the highest point of its trajectory (262b25ff.; 263a4f.). Galileo (quoted from Drake-Drabkin, *On Motion and Mechanics*, Madison 1960, p. 96) criticized the result by criticizing what Aristotle says when introducing it. There is a temporary halt, says Aristotle, 'for one point must be reckoned as

two, being the finishing point of one half [of the motion] and the starting point of the other' (262b23ff.). Galileo objects that, though the turning point may be *described* in two different ways, as the starting point of one segment and as the finishing point of another, it still *is* only one point that corresponds to one instant, the instant of reversal. But quite apart from the fact that Aristotle demands an interval (the moving object 'cannot have arrived at [a certain point] and departed from [it] simultaneously for in that case it would simultaneously be there and not there at the same moment; so there are two points of time concerned with a period of time between them' [262b28ff.]), an interval is demanded also by his general account of the difference between mathematical and physical entities (cf. the quotation at the end of 5).

Galileo also uses an example to ridicule Aristotle's account: A line *ab* moves toward *b*, with the movement gradually becoming slower. A body *c* situated on the line moves toward *a*, with the movement gradually becoming faster (Figure 3).

> Now it is clear that in the beginning *c* will move in the same direction as the line . . . . And yet since the motion of *c* is faster, at some moment *c* will actually move towards the left and will thus make a change from rightward to leftward motion over the same line. And yet it will not be at rest for any interval of time at the point where the change occurs. And the reason for this is that it cannot be at rest unless the line moves to the right at the same speed as body *c* moves to the left. But it will never happen that this equality will continue over any interval of time, since the speed of one motion is continually diminished and the other continually increased.

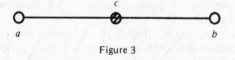

Figure 3

This remark criticizes an argument (motions can be subdivided only by introducing physical changes into them) leading to an assertion (temporary halt at points of reversal) by presenting a case that fits the argument; for it is clear that, if the argument is correct, i.e., if motions can be marked (subdivided) only by introducing physical changes such as halts, then the reversal of *c* will imply a temporary halt of *c* and, thereby, temporary halts of the two processes of acceleration that create it.

12. Different physical entities (and different mathematical entities) may by separation yield the same mathematical entities, e.g., lines and areas, but this does not mean that they can be compared. Thus, curved angles and straight angles can be mapped (Figure 4) on to the linear continuum (or, using Aristotelian terminology, a linear continuum 'can be separated' from both of them), but there is no way of saying that a given curved angle is smaller, equal to, or larger than a straight angle (there is no way of inserting a straight angle into a curved angle [Euclid, *Elements* iii, 16]. Similarly, the area of a circle cannot be equal to, smaller than, or larger than the area of a polygon. Bryson's attempt to measure the area of a circle by the area of a polygon (the circle is smaller than any circumscribed polygon and larger than any inscribed polygon; things smaller or larger than the same are equal to each other; hence, there is a polygon equal in area to the circle) was criticized by Aristotle for precisely this reason. 'The equal is what is neither great nor small but could be great or small *because of its nature*' (1056a23ff.). According to Aristotle, Bryson uses 'a common middle term' (*Anal. Post.* 75b42f.) – 'area' – referring to an entity that has been separated both from the circle and from the polygon but without inquiring whether before the separation this entity occurred together with further properties, different in both cases and preventing a comparison: 'he does not deal with the subject matter concerned' (*Soph. Ref.* 171b17f.), viz., the area *of a circle*. (The geometer, says Aristotle, need not even consider Antiphon's 'exhaustion' of the circle by polygons. The procedure is not just mistaken, it misses its subject matter [185a16f.].)

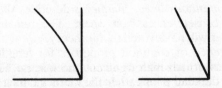

Figure 4

13. Considerations such as these may explain why Aristotle refused to measure qualitative change by length and why he thought that linear and circular motions were incomparable (227b15ff.; 248a10ff.). They also explain why the Euclidian definition of mathematical proportions (*Elements* V, Def. 3) is explicitly restricted to 'homogeneous' quantities and why Greek mathematicians and later mathematicians up to and including Galileo never introduced 'mixed' quantities such as velocity defined as a quotient of spatial and temporal magnitudes. But the fact that certain entities such as areas can be separated both from circles and from polygons indicates that they share some properties and that general statements can be made about them. According to Aristotle, such general statements play an important role in mathematics: 'there are certain general mathematical statements which are not restricted to [special] substances' (1077a9ff.). For example:

> The exchangeability of the inner terms of a proportion was in earlier times proved separately for numbers, lengths, bodies and time while this can be shown for all by a single proof. But for lack of a unified notation and because numbers, lengths, times and bodies looked so different each of them was grasped separately. Now the proof is general for what it asserts is true not only for lengths as such or numbers as such but for what is assumed to hold for the whole (*An. Pr.* 74a17ff.).

Similarly, some principles are valid for various sciences. An example is the principle that equals taken from equals give equals (*An. Post.* 76a38ff.). But the generality cannot be taken for granted and must be ascertained by special arguments.

14. Aristotle offers such arguments for linear extension, time, and motion. Linear extension, time, and motion differ in many respects. They are not 'homogeneous' in the sense of Euclid V, Def. 3 (see 12). Yet, they have common properties. *Aristotle's theory of continuous linear manifolds* describes the properties and draws consequences from them. The remainder of the present paper will be devoted to this theory.

The existence of common properties for length, time and movements is already hinted at by common sense. For example, 'we say that the road is long when the journey is long and that the journey is long when the road is long – the time, too [is called long] if the movement and the movement if the time' (220b29ff.). We also notice that the commonsensical distinction between what is 'in front' and what is 'behind' is a distinction that applies to place and, hence, to extension and that 'it must also be true of movement, for the two [extension and movement] correspond to each other. But if so, then also for time [before and after] for time and movement always correspond to each other' (219a15ff.; cf. 218b21ff.). For Aristotle such analogies are 'reasonable' (220b24), because extension, time, and movement are all *continuous* and *divisible quantities* (24ff.) and are related to each other in such a way that whatever is true of one is true of all (231b19ff.). Now 'continuous', 'divisible', and 'quantity' as defined in geometry are technical terms. Moreover, Aristotle has a rather elaborate theory of continuity. It needs, therefore, special arguments to show that the analogies noticed apply to these technical entities as well, and to determine their limitations.

15. 'By continuous,' defines Aristotle, 'I mean what is divisible into further divisibles without end' (232b25f.), the parts being 'distinguishable from each other by their place' (231b6f.).

The second part of this definition (and of the arguments that surround it) restrict the discussion to linear extended continua. Other types of continua such as sounds (226b29) and further properties not related to place (31f.) are mentioned but not examined. The definition makes assumptions that may seem obvious to modern readers but that require analysis and were not regarded as trivial in Aristotle's own time. The assumptions are (1) that there are entities that can be divided at any point and in any interval, however small; (2) that the division does not change the extension of the entities and of any one of their parts; and (3) that the divison does not obliterate any interval, however small.

Assumption 1 was opposed by mathematicians (Democritus perhaps included), who assumed minimal lengths or 'indivisible lines'. Assumptions 2 and 3 were criticized by Zeno, who denied the existence of things without thickness, mass, and extension. 'For that which neither when added makes a thing greater nor, when subtracted, makes it less, he asserts to have no being – evidently assuming that whatever has being is a spatial magnitude' (1001b6ff.). During the eighteenth and nineteenth centuries, assumption 1 was often supported by reference to something called 'intuition' (*Anschauung*), and the idea of a continuum as a kind of 'gooey substance out of which we pick a point here and a point there' (H. Weyl, 'Über die neue Grundlagenkrise der Mathematik') was referred to this dubious source. Are there better ways of supporting the idea of a linear continuum and of the assumptions involved?

16. One of the arguments against indivisible lines was thay they provide a common measure for all lengths so that there would not be any incommensurable magnitudes. Another argument was that the laws of geometry would cease to be valid below a certain length: the line bisecting the angle of an isosceles triangle consisting of minimal sides could no longer be said to hit the middle of the opposing side (*On Indivisible Lines*, 970aff.). To see these criticisms in the proper light, let us consider an important consequence of incommensurability.

One way of finding the greatest common measure for two magnitudes was the method of *antanairesis*: subtract the smaller from the larger, the difference from the smaller, and so on until you obtain zero (Figure 5). The last number in this sequence before you reach zero is the sought after measure. The procedure was used by mathematicians, but it was also used by carpenters, architects, and geographers for finding the greatest common measure for physical lengths.

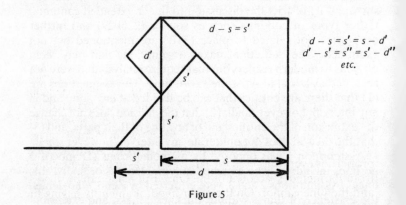

Figure 5

For incommensurable lines such as the side and the diagonal of a square, *antanairesis* has no end. According to some authors, such as Kurt von Fritz, incommensurability was discovered by discovering this property.

Incommensurability could be discovered in this manner only by people who took it for granted that geometrical relations are independent of the size of the figures described. But this was a matter that the Pythagoreans did *not* take for granted. They assumed that the void was structured by indivisible units kept separated by the void. On such an account geometrical relations cease to be valid below a certain miminal length, and the discovery cannot be made. This is a strong argument for giving precedence to the proof reproduced in Euclid's *Elements* Book X: Assume that the relation between the diagonal D of a square and its side S can be expressed with the help of integers, d and s. Then $d^2 = 2s^2$. Reduced to the lowest form, this means that d^2 is even; therefore, d is even and s if odd. Now, since d is even, $d = 2f$ and $2f^2 = s^2$ or, reduced to lowest form, s is even. Hence, s is both even and odd.

According to Eudemus (report in Pappus, commentary of Euclid, i.44), the Pythagoreans founded not only arithmetic but also geometrical algebra. It is often assumed that they did so in order to treat incommensurables in a rigorous manner: numbers cannot do the job, so lines are introduced in their place. But, if that was the motivation, then the conception of a line must have undergone a drastic change – from a collection of individual units separated by a void to a genuine continuum whose parts, however small, have exactly the same structure as the whole. Was this transition the result of a discovery made inside the Pythagorean school, or was it due to outside ideas?

There did exist an outside view that contained all the elements of a continuum – Parmenides' view of the One. According to Parmenides (Diels-Kranz, *Fragmente der Vorsokratiker*, Berlin, many editions, B8, 29ff.), being is 'in its totality homogeneous (*homoion*), it is nowhere more or less' and it is 'connected as a whole'. The word Parmenides uses for 'connected', *xynechés*, is the technical term used by Aristotle in his own account. I suggest that the idea of a line as a continuous entity with the same properties everywhere, in the large as well as in the small, comes from Parmenides' account of the One. But a line can be divided and the One cannot. On the other hand, if an entity having the properties of the One can be divided (note that the division must come from the outside – it is not part of the line itself), then, if it can be divided at one place, it can be divided everywhere, and in exactly the same manner, because of homogeneity. Considerations such as these may give us some historical understanding of assumption 1 of (15). The assumption was not trivial, and it was not based on intuition.

17. Assumptions 2 and 3 can be supported by arguments showing that the entities used for subdividing a linear continuum are unextended and indivisible. Aristotle offers such arguments for the case of the now, i.e., the instant that divides time into past and future. And, since he shows that time, extension, and loco-motion are three different but structurally similar linear con-tinua, this argument applies to all subdivisions.

To be shown: The primary now is indivisible (a moment or interval for an event or a change is 'primary' if it does not contain any interval in which the event or the change does not occur [235b34ff.]; thus, the statement that Caesar was assassinated in the year 44 B.C. does not give the primary or immediate time for the event).

Argument (233b33ff.): The now is a limit of the past, for no part of the future lies this side of it. It is also a limit for the future, for no part of the past lies beyond it. Both limits must coincide. If they were different, then they would either include time or there would be no time between them. If there is no time between them, then they succeed each other, but they cannot succeed each other for succession assumes separation (this point will be argued below in 21). If there is time between them, then it can be divided: part of it will be in the future, part of it will be in the past – we are not dealing with the primary now. Result: The primary now cannot be divided (and, being a limit of an extended continuum, is also without extension).

Leibniz, in a short essay on motion and continuity (*Philo-sophische Schriften*, ed. C. I. Gerhard, Berlin 1885-1890, Vol. 4, pp. 228f., esp.§4) that contains a systematic presentation of part of Aristotle's theory of motion and continuity, has improved this argument and extended it to all limits, divisions, and termina-tions. Take a line *AB* and consider its beginning, *A* (Figure 6). Divide the line in the middle, at *C*. *CB* does not contain *A*. *AB* is, therefore, not the 'primary' end, and *CB* can be omitted. Divide *AC* at *D*. *CD* does not contain the end; hence, *AC* is not the primary end and *DC* can be omitted – and so on as long as we are dealing with any interval, however small. Conclusion: The primary end of the line *AB* (or the primary division of a line that extends beyond *A* to the left) is indivisible (cf. also Euclid, *Elements* i, Defs. 1 and 3). This establishes assumptions 2 and 3 of 15.

18. Another argument consists in pointing out that divisions, ends, and sections do not belong to the same category as do lines: 'the now is not time but an accidental property of it' (220a21f.) and 'lines, or the things derived from them [such as points] are not independently existing substances but sections and divi-

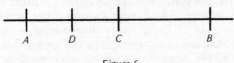

Figure 6

sions . . . and limits [of something else] . . . and they inhere in something else' (1060b10f.); however, they adhere 'not as parts' (220a16) that could exist independently (1060b17) and could therefore obliterate a corresponding interval of a line when dividing it. Zeno's objection (1001b5f.; cf. 14) is met by noting that adding limits and/or subdivisions increases number (of subdivisions), not size. For example, a thrice-divided line is not longer than a twice-divided line, but it has different properties.

19.   I am now ready to present Aristotle's theory of continuity and motion. I will not explore all the ramifications of the theory, nor will I try to remove all the lacunae and ambiguities (there are very few of them). I will certainly not try to present the theory in a form that satisfies modern standards of mathematical rigour. To start with, there are no generally accepted versions of such standards – creative mathematicians, physicists, and systematizers have always gone different ways. Second, standards of rigour, when firmly imposed, often inhibit discoveries or make the formulation of the discoveries impossible (cf. the views of Lakatos and, to a lesser extent, of Polya). Third, dressing up Aristotle in modern garb would conceal his achievements. Aristotle was the foremost mathematical philosopher of his time, and he was well acquainted both with the technical problems and with the most precise ways of formulating them. Trying to present him in modern terms would disrupt this historical connection. Fourth, those modern thinkers who either criticize Aristotle (Galileo, for example; cf. 25) or who repeat him (H. Weyl; cf. 21) use language very similar to his own.

20.   The theory is based on a series of *definitions* (226b18f.): Things are *together* when they have the same primary place (for 'primary', see 17; for place, cf. 6); they are *apart* when this is not the case. Things *touch* when their extremes are together. *A is contiguous with B* if *A* touches *B*. *A is continuous with B* if *A* is contiguous with *B* and the ends of *A* and *B* are one 'or, as the word implies, are contained in each other' (227a15f.). *Between* is defined by reference to change (as are other notions in Aristotle's physics; cf. 262aff.). Every change involves opposites (cf. 190b34ff.) that are contraries (227a7ff.) and as such extremes (226b26). Every stage of a continuous motion with given extremes that has passed one extreme and not yet reached the other is *between* these extremes (the extremes need not be locations; they can be sounds, colours, and other properties that admit of a linear arrangement). *A is the successor of B* if *A* and *B* are of the same kind and if there is nothing of this kind between *A* and *B*. *A* and *B* are part of a *linear continuum* if there are continua *C, C', C'', C''' . . . Cⁿ* between *A* and *B* such that *A* is continuous with *C, C* is continuous with *C' . . . ,* and *Cₙ* is continuous with *B*. From now on, the discussion will be restricted to linear continuous manifolds in this sense.

Aristotle defines 'between' after 'touching', thus assuming continuity before it has been defined. I have changed the sequence, using the definition of 'betweeness' to restrict the earlier definitions to series of events. The definitions make it clear that 'continuity belongs to things which become one by touching' (227a14ff.). They thereby solve the problem as to what it is that makes a continuum such as an individual line a single individual thing (cf. 1077a21ff.). In the physical world, things become one by virtue of a functional unit or a soul – otherwise, they are aggregates. Linear continua hold together because their parts are connected in the manner just described.

21. The definitions imply:

Proposition 1: Linear continuous manifolds do not contain (do not consist of) indivisibles.

Proof: Indivisibles have no parts, therefore they have no ends and cannot hang together in the way described.

For example, lines do not *actually* contain points though, being divisible anywhere, they contain points *potentially*. And, since points mark intervals, we must also say that the parts of a line, such as its right half or its second fifth from the left, are contained in the line only potentially, not actually. The line is one, whole, and undivided until its internal coherence is interrupted by a cut.

Galileo (*Two New Sciences*, quoted from Stillman Drake's

translation, London 1974, 42f.) ridicules this idea in the following manner:

> SALVIATI: . . . I ask you to tell me boldly whether in your opinion the quantified parts of the continuum are finite or infinitely many.
> SIMPLICIO: I reply to you that they are both infinitely many and finite; infinitely many before division, but actually finite [in number] after they are divided. For parts are not understood to be *actually* in their whole until after [they are] divided, or at least marked. Otherwise they are said to be *potentially* there.
> SALV.: So that a line, 20 spans long, for instance, is not said to contain twenty lines of one span each, actually, until after its division into twenty equal parts. Before this, it is said to contain these only potentially. Well, have this as you please and tell me whether, the actual division of such parts having been made, that original whole has increased, diminished, or remains still of the same magnitude?
> SIMP.: It neither increases or diminishes.
> SALV.: So I think, too. Therefore the quantified parts in a continuum whether potentially or actually there, do not make it quantitatively greater or less. . . . .

The implication of this little dialogue is that the lack of effect upon size makes the distinction between actual parts and potential parts meaningless. Stillman Drake, the translator and commentator, agrees (p. 42. fn. 27): 'Here Galileo proceeds to show the distinction is meaningless mathematically unless it affects quantity or magnitude.' But a linear continuum in Aristotle's sense not only has extension, it has a structure – and this structure is changed by every division (using a similarly superficial argument, one might say that there is no distinction between a litre of wine and a litre of water because both have the same volume). The objection that Galileo is speaking as a mathematician does not remove the difficulty, for it is just Aristotle's point that mathematical entities, apart from being extended, have also a structure; otherwise, there would be no difference between the number five and a line five inches long.

Proposition 1 implies:

Proposition 2: Linear continuous manifolds (LCMs for short) are divisible into LCMs without limit and, further, that (231b5ff.):

Proposition 3: No point of an LCM can be a successor of another point of an LCM (for that would

assume that the line between the two points
cannot be further divided).

Propositions 1, 2, and 3 express an idea that is similar to the
modern concept of an everywhere dense manifold. The difference is that the modern idea assumes the points as given while for
Aristotle they are potential and have to be actualized by subdivision.

22. An undivided motion is an individual whole, without parts,
and it is completed in a single step. The same is true of the
distance covered and of the time it needs for the motion to occur.
Dividing the motion means dividing the time and the distance;
dividing the distance means dividing the motion and the time.
We may conjecture that linear extension and time are LCMs just
as motion is. Given the continuity of motion, length, and time,
we can introduce a definition of 'quicker' that was accepted in
antiquity and was still used by Galileo: Quicker is what either
covers a greater distance in the same time or the same distance in
a smaller time or a greater distance in a smaller time (232a23ff.).
This is longer and much more clumsy than the corresponding
modern definition. The 'clumsiness' is intentional: Distance and
time may have some abstract properties in common (continuity,
divisibility), but they are not 'homogeneous' magnitudes (13).
Therefore, they can only be related to themselves, distance to
distance, time to time, motion to motion.

Now consider two objects, one quicker, the other slower.
Assuming that any motion can take place within any period of
time (232b21f.) and that for any period of time there can be a
distinction between faster and slower (233b19f.), we see (Figures
7 and 8 and 233a8ff.) that the faster will subdivide the time and
the slower, the distance, thus establishing this: if length is continuous, so is time, and vice versa.

'This conclusion follows not only from what has just been said
but also from the argument that the opposite assumption implies
the divisibility of the indivisible' (233b16ff.). Assume (since
velocities can stand in any relation) that one body covers the
distance AB while another body covers two-thirds of that
distance in the same time (Figure 9). Let the intervals $Aa = ab =
bB$ be indivisible, and let the same be true of the corresponding
times; i.e., let $Rd = de = eS$. Then the slower body when arriving
at the possible division a will divide the time at f, thus dividing the
indivisible.

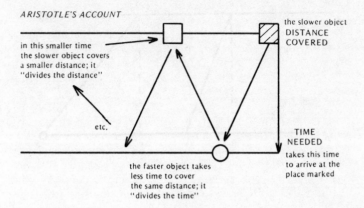

ARISTOTLE'S ACCOUNT

Figure 7

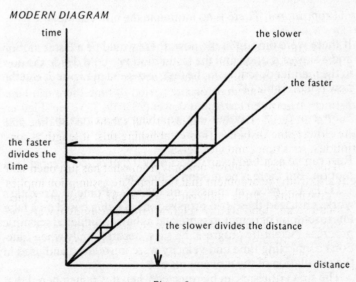

MODERN DIAGRAM

Figure 8

238

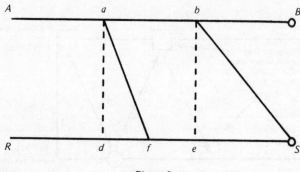

Figure 9

This supports the conjecture and shows the coherence of the notions used and the connections established. But there is no absolute proof of these notions and these connections.

23. Next, we have a sequence of theorems concerning the relation of time to motion and distance.

Propositon 4: There is no motion in the now (234a24).

If there were motion in the now, there would be a faster motion and a slower motion and the faster motion would divide the now in the manner described in the text discussion of Figure 7. But the now is indivisible (17).

Propositon 5: Neither can anything be at rest in the now (234a33f.).

Rest can be ascribed to an object only if that object is capable of motion. But there is no motion in the now.

G. E. L. Owen, in an essay on the role of time in Aristotle's work, criticized these two propositions as being based on a false interpretation of common sense and as having hindered scientific progress. Scientific progress, he says, occurred only when functions connecting time and velocity were introduced and used in the calculation of the movements of bodies.

The first criticism can be rejected when it is remembered that Aristotle's notion of continuity owes a debt to Parmenides (16).

It is true that Aristotle occasionally uses homely examples such as glue and nails (227a17) to *illustrate* continuity. But the *content* of the notion lies in its consequences, and among those consequences we have Proposition 2 (limitless divisibility), which can be proved only via a postulate of homogeneity such as the one proposed by Parmenides.

The second criticism only shows that scientists can get very far by thinking very little. In 21, I quoted Galileo to the effect that what mattered was the length of a line, not its structure. This attitude served scientists well as long as the problems they encountered did not involve structure. Trouble arose in quantum mechanics, where considerations of structure became essential. In the attempt to resolve the problems connected with structure, physicists introduced ideas very similar to those expressed in Propositions 4 and 5 (uncertainty relations between time and energy) and can now be said to agree with Aristotle's principle that 'the motion of what is in motion and the rest of what is in rest must take time' (234b9f.).

Proposition 6: A point has no place (212b24f.).

This follows from the definition of place in 6 as the spatial correlate of Propositions 4 and 5. The three propositions (and some others that are about to follow) are anticipated in Plato's discussion of the One, *Parmenides* 137ff. This passage also contains material Aristotle seems to have used in his definition of continuity (see 20).

24. According to 9, motion can be subdivided only by being brought to a temporary halt – the 'mobile must stop and begin to move again' (262a24f.). When the motion stops, the moving object is in a well-defined place and has well-defined properties: e.g., an object is grey when it was stopped on the way from white to black (234b18f). Being in a well-defined place and having well-defined properties is the character of an object that does not move. Moving, therefore, implies not being in a well-defined place and not having well-defined properties. It seems that Aristotle, while occasionally drawing such a conclusion (cf. Proposition 14), was not always prepared to take this step (cf. the restriction: 'for as a whole [the object] cannot be in both [the initial and the subsequent state of change] or in neither' [234b17]). He concludes, rather, that 'during the whole process of change [the object] must be partly under one condition and partly under another'; i.e., it must be divided into parts that are in one condition and parts that are in another condition.

Proposition 7: What changes is divisible (234b10).

Applied to locomotion, this means that we are dealing with elastic and deformable objects. Note the similarity to the relativistic account of the motion of extended objects.

Proposition 8: Whenever a change is completed, the changing thing is in the state into which it has changed (literally: what has changed is that into which it has changed [235b6ff.]).

This simply follows from the meanings of the terms.

Proposition 9: The primary time when what has changed has completed its change is indivisible.

Assume it is divisible, and change occurs in both parts. Then the change has not yet come to an end. If change occurs in one part only, then we are not dealing with the primary time (235b33ff.).

Propositions 8 and 9 are connected with the fact, emphasized in 11, that clearly marked stages of a motion are accompanied by an interruption of that motion. Now motion involves opposites (189a10); one of the opposites is the 'for the sake of which' or the aim (*telos*) of the motion (194b33). When the aim is reached, the motion is completed and, therefore, interrupted; and the interruption provides an indivisible limit (236a13) for the motion. Propositions 8 and 9 express this situation.

Proposition 10: There is no indivisible primary time when what changes is completing its change (239a1ff.).

The reason is that what is completing its change is in motion and (Proposition 4) there is no motion in an indivisible moment. In a similar manner, Proposition 5 yields Proposition 11.

Proposition 11: Neither is there an indivisible primary time in which rest first occurs (239a10).

Further:

Proposition 12: Everything that changes at a certain time changed before that time.

Assume *AB* to be the primary time for the change. The change must occur at a point *a*, in between *A* and *B*, but also at *b*, in between *A* and *a*, and at *c*, in between *b* and *A*, and so on. Therefore:

Proposition 13: There is no such thing as the beginning of a process of change (236a13f.).

25. The results of the preceding section may be summarised as follows. Every change is characterized by a well-defined indivisible moment, the primary moment when the change has been completed. There is no last moment when the change is still on, there is no first moment when the change starts, and there is no first moment of rest after the change has been completed.

One might be inclined to regard this situation as a trivial consequence of the fact that a series of changes terminating in an aim is closed on the right and open on the left and that it is everywhere dense (cf. chapter 4 of E. V. Huntington. *The Continuum*, Cambridge 1917). But the comparison is misleading in various respects. To start with, the structure of an Aristotelian line differs from the structure of a dense series. The elements of a dense series all exist and constitute the series. An Aristotelian line, on the other hand, is one and undivided until the parts are actualized by special means. Second, the end point of the change is actual not because all points of a change are actual and not because the change has been stopped by external means, but because of the particular way in which the change, every change, is completed: it arises because of the inner structure of the process of change. Third, it has no beginning because there is no motion in the now.

The difference between the Aristotelian continuum and the mathematical continuum has been described with great clarity (but without reference to Aristotle) by Hermann Weyl in his essay *Das Kontinuum* (Leipzig 1919, p. 71):

> There is no agreement between the intuitive continuum [which is how Weyl describes the continuum viewed as an indivisible whole] and the mathematical continnum [which consists of points] . . . ; both are separated by an unbridgeable gulf. However we have reasonable motives which, in our attempt to comprehend nature, make us

move from the one to the other. They are the same motives that directed us from the world of human experience in which we live our normal lives towards a 'truly objective', precise and quantitative physical world 'behind' experience and that made us replace the colour qualities of visible objects by aether vibrations . . . . Our attempt to build up analysis [from indivisible units] may therefore be regarded as a *theory of the continuum* which has to be tested by experiment just as any other physical theory.

The mathematical reconstruction of the continuum, Weyl says at a different place ('Über die neue Grundlagenkrise der Mathematik,' *Math. Zs.* Vol. 10, 1919, p. 42),

selects from the flowing goo . . . a heap of individual points. The continuum is smashed into isolated elements and the interconnectedness of all its parts replaced by certain relations between the isolated elements. When doing Euclidian geometry it suffices to use the system of points whose coordinates are Euclidian numbers. The continuous 'space-sauce' that flows between them does not appear.

This is Galileo's attitude (see the quotation in 21), except that Weyl is aware of the loss and of the possibility that its consequences may turn up in physics. The 'continuous space sauce' may make itself noticed as we proceed into new domains of research. There are some physicists who think that it already appeared in microphysics.

26.   According to Propositions 4 and 5, there is neither motion nor rest in an instant; every motion fills an interval of time. The location of an object that moves in space is indeterminate in accordance with the size of this interval. If location is indeterminate, then so is length.

Proposition 14: What moves has no well-defined length [in the direction of the motion].

Conversely, a definite length can be assigned to an object only if it can be made to 'cover' a stationary measuring stick, i.e., if it is at rest. Being at rest takes time (Proposition 5); hence:

Proposition 15: An object can be said to have a well-defined length only if it is at rest for an interval of time, however small.

27. I conclude with a brief account of the way in which Aristotle solves Zeno's paradoxes of motion. Aristotle describes four such paradoxes (239b10ff.). His description is the earliest detailed account we have of Zeno's arguments.

According to the first paradox, motion is impossible because before getting to a point the moving thing must first cover half the distance, then half of that half, and so on. One solution has been described in 9. Aristotle has a second solution, which he regards as less satisfactory (263a4ff.).

According to the second paradox, the 'Achilles', the faster can never overtake the slower 'since the pursuer must first reach the point where the pursued started, so that the slower must always hold a lead.' This Aristotle regards as a different version of the first paradox, and he solves it in the same manner.

The third paradox, the 'arrow', states that, as a flying arrow at any moment of its journey occupies a place equal to its own dimensions, it is at rest at any moment of its journey and therefore at rest throughout the journey. The paradox is solved by reference to Propositions 4 and 14.

The fourth paradox, which is more difficult to interpret, is supposed to show 'that half the time is twice the time'. There are (Figure 10) three series of masses, *A, B* and *C. A* is at rest; and *B* moves to the right, *C* to the left, both with equal speed. While *C* passes all the *B*s, *B* passes only half of the *A*s. Assuming that passage of two masses takes the same time whether the masses are moving or at rest, we can say that *B* while passing half the *A*s passes all the *C*s and therefore takes half the time, while the *C*s pass twice as many *B*s as *A*s and so take twice the time for the same process. Aristotle denies the assumption and so removes the paradox.

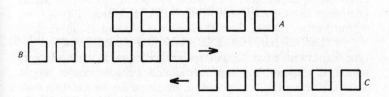

Figure 10

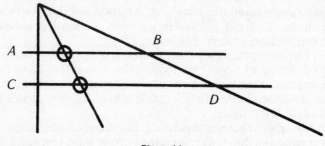

Figure 11

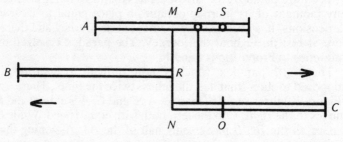

Figure 12

Rafael Ferber, in an interesting and provocative book (*Zenons Paradoxien der Bewegung*, Munich 1981), has suggested a connection between this paradox and some earlier versions of the idea that what is infinitely divisible has the same number of indivisibles no matter what the size. Today, this idea is illustrated by drawings such as Figure 11. To every point of *AB*, there corresponds one and only one point of *CD*, and the two lines have an equal number of points. According to Arpad Szábo (*Anfänge der Griechischen Mathematik*, Budapest 1969), from whom Ferber quotes, Euclid *Elements* i, Axiom 8: the whole is bigger than the part (which in proofs is replaced by the stereotype 'otherwise what is smaller would be equal to what is larger, which is impossible') was introduced because some people denied it (no other reason can be imagined for formulating such an obvious principle). One person to deny the principle was Anaxagoras (Diels-Kranz, B3).

For any small thing there are things which are still smaller – for it is

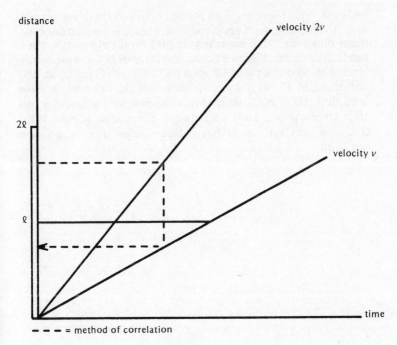

distance

velocity 2*v*

velocity *v*

2ℓ

ℓ

time

- - - = method of correlation

Figure 13

impossible that being ceases. But also for what is big there is always something bigger *and it is equal to the small in amount.* Taken for itself, however, everything is both big and small.

We may assume that this assertion underlying Anaxagoras' idea that every piece of material contains ingredients of everything else – there is flesh in metal, metal in air, air in bone, and so on – is also connected with Parmenides' assumption of the homogeneity of the One, introduced in 16 above. If the One is homogeneous throughout, then the smallest part has exactly the same structure as the whole; e.g., it has the same number of parts (subdivisions).[2] Is it possible to find an interpretation of the fourth paradox that leads to the same result? It is possible! (see Figure 12). Take any point of *C*, assumed to be continuous. When this point, say *O*, passes the right end *R* of *B*, then *R* will be

---

2. Of course, Parmenides' One has no parts and, therefore, no structure apart from being homogeneous. But, if we contradict Parmenides and add such a structure, the homogeneity will guarantee its presence in all parts.

beneath P, half way between R and O, and so P will correspond to O. Conversely, to every point S on A there is one and only one point on C, viz., that point that is $2MS$ on the right of N. On a modern diagram (Figure 13), the correlations become especially clear: the whole is mapped upon half of itself. Considering the definition in 15, we may conjecture that Aristotle might have accepted this result, *provided* that the mapping is between cuts that create points and not between preexisting points. What consequences he would have drawn from it is a different question.

9

Galileo and the Tyranny of Truth

This was a talk given, by tape, to the Pontifical Academy in Cracow, Poland. It has been largely rewritten, but I have retained the original style. A bowdlerized version of the talk was published in The Galileo Affair: A Meeting of Faith and Science, *edited by C. V. Coyne, M. Heller and J. Zycinski, Vatican City 1985.*

Esteemed members of this conference! First let me apologize for not coming in person but sending an electronic representative instead. This representative – the tape I am preparing for you right now – will give you some of my opinions on the topic of the conference but unfortunately it won't be able to reply to your questions or to answer your objections. The reason for my absence is a series of accidents which prevented me from getting my visa in time and so I am sitting here, all by myself, in a small room, on top of a large building, with a view of the Swiss Alps, some green hills and Lake Zürich right in front of me. The landscape is wonderful but, alas, there is not a single human face – which is somewhat of a disadvantage when you are supposed to be giving a talk. Still, I hope I will be able to make my presentation lively and interesting and that I will also be able to anticipate which parts of my lecture you might receive not with a smile, nor with attentive curiosity, but with frowning disbelief: those parts, naturally, I shall treat with special care.

My second apology is about the way in which I am going to present my opinions. As far as I am concerned the best way of describing a historical conflict is to introduce the *individuals* that created it, to describe their temperament, their interests, their hopes and ambitions, the information at their disposal, their social background, the individuals and institutions they felt loyal to and that supported them in turn, and many similar things. Then one would have to explain how the individuals got entangled with each other and with the associated institutions, how they viewed the entanglement and how they reacted to it; one would have to explain, for example, how they used the power they possessed to reduce the entanglement and to make it serve their purpose, one would have to describe how the conflict was shaped by the legal and social laws of the time and the tensions between those laws and the temperament of the individuals – and so on.

This, to my mind, would be the most desirable way of introducing a conflict such as the conflict between Galileo and the Church but, alas, it can't be done. The material is much too voluminous to be surveyed in an hour and, besides, I know only a small part of it. I shall therefore use a different procedure to make my point: I shall 'rise to a higher level of abstraction'. I shall not talk about individuals and their idiosyncracies, I shall talk about *traditions*. I shall present the conflict between Galileo and the Church as a conflict between traditions and I shall try to show that the tradition represented by the Church had interesting ancestors in antiquity and has progressive defenders today. Of course, speaking of traditions (or paradigms, or research programmes, or themes, to use terms from more narrow fields than that before us) is a natural approach in history, sociology, and philosophy. I use the approach not because I am in love with it, but because of the difficulties just mentioned. So don't forget that we shall always be at least two steps removed from reality.

The traditions I have in mind are traditions concerning the role of experts in society. In earlier writings, I have described two such traditions. One regards an expert as the final authority on the use and interpretations of expert views and expert procedures, the other subjects the pronouncements of experts to a higher court which may consist either of superexperts – this was Plato's view – or of all citizens – this seems to have been recommended by Protagoras. I suggest that the opposition between Galileo and the Church was analogous to the opposition between

---

1. See chapter 1, section 6, of the present volume.

what I have called the first and the second view (or tradition). Galileo was an expert in a special domain comprising mathematics and astronomy. In the classification of the time he was a mathematician and a philosopher. Galileo asserted that astronomical matters should be left to astronomers entirely. Only 'those few who deserve to be separated from the herd' could be expected to find the correct sense of Bible passages dealing with astronomical matters, as he wrote in his letter to Castelli of December 14, 1613 (Copernicus before him and Spinoza afterwards used similar language; this is an old topic, as Hanns-Dieter Voigtländer, *Der Philosoph und die Vielen*, Wiesbaden 1980 shows: it occurs already in antiquity). In addition Galileo demanded that the views of astronomers be made part of public knowledge in exactly the form in which they had arisen in astronomy. Galileo did not simply ask for the freedom to publish his results, he wanted to impose them on others. In this respect he was as pushy and totalitarian as many modern prophets of science – and as uninformed. He simply took it for granted that the special and very restricted methods of astronomers (and of those physicists who followed their lead) were the correct way of getting access to Truth and Reality. He was a perfect representative of what I have called the first view or tradition.

The position of the Church, on the other hand, was very similar to the second view (in its Platonic, not its Protagorean version). Astronomical knowledge, according to the Church, was important and interesting, and it was actively pursued by some of its members. But the models which the astronomers produced to account, say, for the paths of the planets could not be related to reality without further ado. They arose from special and limited purposes and all one could say was that they served these purposes, viz., prediction.

Exactly this point is made in the first part of a famous letter which Cardinal Bellarmino, master of controversial questions at the Collegio Romano, wrote to Paolo Antonio Foscarini, a Carmelite monk from Naples, who had inquired about the reality of the Copernican System. The letter is often quoted, and even more often criticized by comparing its assertions with certain abstract principles which allegedly govern the practice of science. It looks very different when compared with this practice itself, as we shall see. In my opinion it is a very wise document and it contains sensible suggestions concerning the position of the sciences in our culture.

Bellarmino writes:

It seems to me that Your Reverence and Signor Galileo act prudently when you content yourselves with speaking hypothetically and not absolutely . . . To say that on the supposition of the Earth's movement and the Sun's quiescence all the celestial appearances are explained better than by the theory of eccentrics and epicycles is to speak with excellent good sense and to run no risk whatsoever. Such a manner of speaking is enough for a mathematician. But to want to affirm that the Sun, in very truth, is at the centre of the universe and only rotates on its axis without going from east to west, is a very dangerous attitude and one calculated not only to arouse all Scholastic philosophers and theologians but also to injure our holy faith by contradicting the Scriptures.

To use modern terms: astronomers are entirely safe when saying that a model has predictive advantages over another model, but they get into trouble when asserting that it is therefore a faithful image of reality. Or, more generally: the fact that a model works does not by itself show that reality is structured like the model.

This sensible idea is an elementary ingredient of scientific practice. Approximations are a commonplace in science. They are used because they facilitate calculations in a restricted domain. Their symmetry-properties frequently differ from those of the underlying theory. Hence, if the theory is assumed to correspond to reality, the approximations cannot correspond to it in the same sense. Theories, on the other hand, are often developed as steps towards a more satisfactory but as yet unknown view. They may be successful but the very purpose for which they were introduced forbids us to draw realistic consequences from them. The older quantum theory is an example – and so was Newton's theory of gravitation, at least as Newton saw it. Even a formally perfect theory with surprising predictive powers may fail when regarded as a direct expression of reality. Schrödinger's wave mechanics illustrates this further point. It was elegant, coherent, easy to handle and it was remarkably successful. Schrödinger inferred that elementary particles were waves. Yet looking at a wider range of phenomena Bohr and his school showed that this interpretation conflicted with important facts (there are also two formal obstacles, viz. the so-called reduction of the wave packet and the fact that the theory is not Lorentz invariant). And now take the best theories of modern physics, general relativity in its most recent form and general quantum mechanics. So far it has proved impossible to merge them into a coherent whole – the one theory makes assertions which are flatly contradicted by the other. Can we still assert that we get a correct description of reality from either of them? We cannot. We can say that both theories are useful approximations but we have no idea what the reality they approximate looks like.

All these examples have immediate application to the case of the Copernican theory whose coherence and partial success were also regarded as signs of a close correspondence to reality, both by its author and by writers such as Rheticus and Mästlin. For the Copernican theory was not the only cosmological view in existence and not even the most general one. Hence its success and the coherence of its structure, taken by themselves, did not yet mean agreement with reality. To show such agreement it was necessary to move to a wider domain.

In modern science the wider domain chosen very often is elementary particle physics. Scientists working in this field agree that chemists, biologists, rheologists etc. may have discovered some interesting regularities but they deny that these regularities are basic features of reality. Some modern biologists similarly look askance at things such as botany and bird-watching, asserting that the only real information about the life process comes from molecular biology. In his search for a way out of the difficulties of early twentieth-century science, Einstein relied on thermodynamics. In all these cases models are compared with basic science and their realistic implications are judged accordingly. What was the wider domain that determined reality for the Church?

According to Bellarmino, the wider domain contained two ingredients, one scientific – philosophy and theology; one religious and to that extent normative – 'our holy faith'.

The first ingredient differed from modern measures of reality (molecular biology, elementary particle physics, cosmology) in content but not in function. Philosophy meant mainly the Aristotelian opus, which included a general theory of change and motion, a theory of the (mathematical and physical) continuum (the theory is described in chapter 8 of the present book), a theory of elements and some considerations concerning the structure of the world. Theology dealt with the same subject matter but viewing it as a creation, not as a self-sufficient system. It was and still is a science, and a very rigorous science at that: textbooks in theology contain long methodological chapters, textbooks in physics do not.

The second ingredient means that scientific results, wrongly interpreted, may injure human beings. This ingredient, too, is still with us. Modern defenders of science frequently warn us that a misleading presentation of scientific results or of basic scientific conflicts may lead to irrationalism and thus injure 'our holy faith' in reason. Conversely opponents of reductionism try to interpret scientific results in a manner that no longer violates their 'holy faith' in the integrity of nature, cultures, and individual human beings. The difference between Bellarmino's version and the modern versions is twofold. Most modern versions have ceased relating their faith (in reason, in the integrity of human beings, etc.) to a divine creator; and the institutional backing of the older warnings was stronger than is the institutional backing of reason or of anti-reductionism today. This last difference, however, reflected not the Church, but the entire age. And let us not forget that many modern rationalists try to increase the power of Reason by increasing the power of the institutions that support it.

The second ingredient, as presented by Bellarmino, further implies that questions of fact and reality depend on questions of value. For a positivist this is an unfamiliar and even repulsive idea, but only because he is not aware of his own normative prejudices. A brief look at the history of the concept of reality reveals what these prejudices are.

In Homer events such as dreams, the actions of gods, or illusions were all regarded as being 'equally real' (I put the phrase in quotation marks because the notion of reality that occurs in it already assumes that some things may not be as real as others). No separation existed between a reality outside human beings and the result of a perceiving and distorting agency within. Simplifying the situation somewhat, we may say that this was so because there was no 'mind': there was no special region, no 'subject' that contained events set off from the rest of the world, and capable of distorting it. Anaximander then based all cosmic processes on the changes of a single substance, the *apeiron*. Dreams, gods, premonitions had no place in this world – they became homeless.

What was to be done with them? The answer was given by Parmenides, who introduced a sharp distinction between two kinds of things and processes: real things on the one side, mere appearances on the other. Reality was constituted by novel procedures that differed both from tradition and from simple observation. Appearances were attributed to an erring mind (thus the mind arose partly as a receptacle for all those things that could not be accomodated in the 'real world'). Accepting some phenomena as real and rejecting others as deceptive therefore meant choosing one tradition over another. This became very clear later on, when Gnostics and Naturalists quarrelled about the reality of matter, and it is apparent even today when some scientists claim to have discovered an ultimate reality (elementary particles and their fields) while others emphasize large-scale laws and regard high energy physics as a rather expensive and complicated version of stamp collecting. Aristotle described the situation in wonderfully simple terms when defending the reality of non-Parmenidean things and processes with the remark that they were essential for life in the city:

> Even if there should exist a single Goodness which is universally predicable of goods, or is capable of separate and independent existence, clearly it could not be achieved or attained by man; *but we are seeking something attainable* (*Nicomachean Ethics*, 1096b32f, my emphasis).

He also emphasized, in his book on the soul, that materialistic, psychological and sociological accounts of the soul, serving different purposes, were all correct in their respective domains. Thus the church was not only on the right track when measuring reality by human concerns but it was considerably more rational than some modern scientists and philosophers who draw a sharp distinction between facts and values and then take it for granted that the only way of arriving at facts and, therefore, reality, is to accept the values of science.

At the time of Galileo one important source for the discussion of human concerns was the Bible – it still retains this position in our own times. The Church, being the foremost guardian and interpreter of the Bible, also made it a boundary condition of reality.[1] Newton, who opposed Catholicism, still took this boundary condition very seriously. Research, according to Newton, must use two sources: the Works of God, the magnificent Universe, and the Word of God, the Bible. Details of the Bible (such as the story of the flood) were used by scientists to support scientific views until well into the nineteenth century (catastrophism).

The Church not only used the Bible as a boundary condition of truth and reality, it also tried to impose it by administrative

---

1. Here the opponents of Galileo may have gone beyond 'the common opinion of the holy fathers' mentioned by Bellarmino later in his letter. This 'common opinion' regarded the Bible as a guide in moral, but not in astronomical matters.

measures. Bellarmino is very clear on this point:

> As you are aware, the Council of Trent forbids the interpretation of
> the Scriptures in a way contrary to the common opinion of the Holy
> Fathers.

It is here that the modern reader, and especially the liberal
epistemologist who is familiar with some abstract desiderata but
has never seen science from nearby, is liable to throw up his
hands in despair. To his mind knowledge has nothing to do with
administration and his heart goes out to poor Galileo who had to
put up with such nonsense. But it is not at all certain that a
modern Galileo would have an easier life.

Assume, for example, that he wanted to test the efficacy of
modern scientific medicine by using control groups of volunteers
treated by alternative methods. Then, in many states of the USA
he would be visited by the police, just as Galileo was. Assume he
wants to teach Evolution and Genesis on equal terms, as two
alternative accounts of the origin of human beings. Then he runs
into another legal restriction, namely the separation of State and
Church which puts strong legal and administrative limits on the
transmission of knowledge-claims. Evolution is to be taught as a
fact, or as a theory dealing with facts, while Genesis can at most
be taught as a belief. (This is an inversion of Bellarmino, who
assigned the determination of basic facts to philosophy and
theology and granted science at most an instrumental role – a
further confirmation of the point, made above, that 'reality' is a
value-laden term and that questions of reality are closely connected with human concerns.)

Our modern Galileo will also find that arguments only rarely
suffice to get an idea accepted and financed. The idea must fit the
ideology of the institute that is supposed to absorb it and must
agree with the ways in which research is done there. And there is
no individual human being to whom he could explain his suggestions and whom he could educate in his own ways of thinking –
there are anonymous committees, often stacked with incompe-
tents who regard their own ignorance as a measure of things.[2]
How can an intelligent person succeed in such circumstances? It
is very difficult. Galileo tried to combine philosophy, astronomy,
mathematics and a variety of subjects which are best characterised as engineering into a single new point of view which also
entailed a new attitude towards Holy Scripture. He was told to
stick to mathematics. A modern physicist or chemist trying to

reform nutrition or medicine faces similar restrictions. A modern scientist who publishes his results in a newspaper or who gives public interviews before he has submitted to the scrutiny of the editorial board of a professional journal or of groups with comparable authority has committed a mortal sin which makes him an outcast for quite some time.

Admittedly the control is not as tight as it was at the time of Galileo and not as universal, but this is the result of a more easygoing attitude towards certain crimes (thieves, for example, are no longer hanged, or mutilated) and not a change of heart as to the nature of the crimes themselves. The administrative restrictions on a modern scientist are certainly comparable to those in force at Galileo's time. But while those of the older restrictions which issued from the Church were available in the form of explicit rules, such as the rules of the Tridentine Council, modern restrictions are often implied, not spelled out in detail. There is much insinuating and hinting, but there is no explicit code one could consult and, perhaps, criticize and improve. Again the procedure of the Church was more straightforward, more honest and certainly more rational.

And now comes a very important point: this straightforward and rational boundary to research was not immovable. This is what Bellarmino says very clearly in the last part of his letter:

> If there were any real proof that the Sun is in the centre of the universe and that the earth is in the third heaven, and that the Sun does not go round the Earth but the Earth round the Sun, then we would have to proceed with great circumspection in explaining passages of Scripture which appear to teach the contrary, and rather admit that we did not understand them than declare an opinion to be false which is proved to be true.

Church doctrine, Bellarmino says here, is a boundary condition for the interpretation of scientific results. But it is not an absolute boundary condition. Research can move it. However, Bellarmino continues,

2. Commenting on the history of the steady-state theory, Fred Hoyle writes (Y. Terzian and E.M. Bilson, eds., *Cosmology and Astrophysics*, Ithaca and London 1982, p. 21): 'Journals accepted papers from observers, giving them only the most cursory refereeing, whereas our own papers [papers by Bondi, Gold and Hoyle] always had a stiff passage, to a point where one became quite worn out with explaining points of mathematics, physics, fact and logic to the obtuse minds who constitute the mysterious anonymous class of referees, doing their work, like owls, in the darkness of the night.'

as for myself, I shall not believe that there are such proofs until they are shown me. Nor is it a proof that, if the Sun be supposed at the centre of the universe and the Earth in the third heaven, everything works out the same as if it were the other way around. In case of doubt we ought not to abandon the interpretation of the sacred text as given by the Holy Fathers.

The idea expressed in the last sentence is today accepted by all high school principals and even by some university presidents – don't introduce a new basis for education until you are sure it is at least as good as the old basis. It is also a reasonable idea. It advises us to make basic education independent of fashions and temporary aberrations. Education is not just ideas. It is textbooks, skills, equipment for demonstrations, laboratories, films, slides, courses for teachers, computer programmes, problems, examinations and so on. When built up in a judicious way it can accomodate fashions, aberrations and alternative views and thereby illuminate the process of scientific research; however, it would be very unwise to rebuild it from top to bottom whenever an adventurous new point of view appears on the horizon. Besides, one would not know how to proceed – there always exist many conflicting fashions, aberrations, suggestions, 'bold ideas'. The Church took that into account. It demanded strong arguments before it would consider changing an important piece of knowledge.

But was Bellarmino perhaps dragging his feet? Did he resist in the face of clear and unambiguous evidence? Or, even worse, was he perhaps not informed about the evidence that existed at the time? This technical question, unfortunately, has become *the* question for many investigators. I shall attack the problem connected with it by asking a different question: what would have been the judgement of modern scientists and philosophers of science had they been transferred into the early seventeenth century and asked the question Bellarmino was asked, namely: What's your opinion of Copernicus?

The answer is that different people would have said different things. Science, like any other enterprise, knows hardliners and it knows more tolerant people. There are scientists who read the success of a theory in small hints and there are other scientists who want more substantial proofs. There are scientists who are satisfied with simplicity and intellectual harmony, there are other scientists who want solid empirical backing. There are scientists who are frightened by inconsistencies in a theory or between theory and experiment and there are other scientists who regard such inconsistencies as natural companions of progress. Michelson and Rutherford never fully accepted relativity, Poincaré, Lorentz and Ehrenfest became doubtful after Kaufmann's experiments, while Planck and Einstein, convinced by its internal symmetry, were more persistent. Sommerfeld was very successful in making the older quantum theory as formidable as classical celestial mechanics while Bohr, despite the successes achieved, thought that Sommerfeld was on the wrong track. Pauling loved confusing his colleagues with conjectures taken from simple model-building while they preferred pondering the intricacies of X-ray photographs.

Who knows what any of them would have said if transported back to Bellarmino's desk? Michelson, presented with Galileo's *telescopic observations*, might have pointed to their internal contradictions (the planets are pulled towards the observer, the fixed stars are pushed away; the inside of the moon shows mountains while its circumference is perfectly smooth) and he might have laughed at the attempt to get physical information from an instrument so little understood. And almost all philosophers of science writing today would have agreed with Bellarmino that Copernicus's case was very weak indeed.[3] The strongest argument mentioned by Copernicus, Rheticus and Mästlin, an argument which convinced Kepler,[4] was the harmony created by the Copernican point of view: for the first time there was an astronomical system and not only a set of calculating devices. But the argument can lead astray, as the case of Schrödinger shows: the correct interpretation of a simple and harmonious point of view may differ considerably from the interpretation suggested by a first and superficial look at it.

Galileo was aware of the problem – why else would he have given such weight to his 'decisive proof', the theory of the tides? Besides, the *elements of mechanics* he succeeded in contriving during his lifetime were quite inadequate for providing a dynamics of the planetary system as described by Copernicus. They might have supported circles, but they made nonsense of the epicycles that were still needed for correct predictions and

---

3. Details on these matters are found in my book *Against Method*, especially in the largely rewritten third German edition, *Wider den Methodenzwang*, Frankfurt 1986.

4. The argument also turned up in the discussion of the present talk at Cracow.

they were inapplicable to Kepler's laws which Galileo did not accept anyhow. An acceptable solution came later, with Newton, and even he needed divine interventions to keep the planetary system in order. Besides, Galileo's views on the relativity of motions were incoherent. Occasionally he asserted the relativity of *all* motion, on other occasions he accepted impetus which assumes a fixed reference system.

Galileo's *basic physics* was even worse. Aristotle had provided a general theory of change, motion and the continuum. The theory dealt with locomotion, qualitative change, generation and corruption, and it gave an account of falling stones as well as of the transmission of information from a teacher to an attentive student. The theory of locomotion was very sophisticated, entailing that an object cannot move and at the same time have a precise position. Galileo restricted himself to locomotion and even here used terms much more simple-minded than those already introduced by Aristotle (Aristotle had made a step towards the quantum theory which regards motion as an indivisible whole, while Galileo moved away from that achievement).[5] As a consequence biologists, physiologists (Harvey!), the founders of the new science of electricity, and bacteriologists continued to use Aristotelian ideas until late into the eighteenth century and to some extent even right into the twentieth century (Prigogine has some very nice things to say about Aristotle). Newton certainly took Aristotle's comments on motion seriously, as can be seen from his manuscripts. Einstein, with his contempt for a 'verification of little effects' and with his uncanny ability to divine future splendours in a present mess, might have taken the side of Copernicus but many other physicists would have thrown up their hands in despair. Bellarmino's judgement, therefore, is an entirely acceptable point of view.

With this I conclude my brief description of the form two ancient traditions had assumed at the time of Galileo. The traditions dealt with the role of science in society.

According to the first tradition, society must adapt to knowledge in the shape presented by the scientists. The tradition was defended by Galileo. It was used more recently by scientists as a basis for 'negotiations' with the Church after Cardinal König of Vienna had suggested a closer cooperation:[6] cooperation, the

5. For a detailed discussion of this point, see chapter 8, above.
6. *Physikalische Blätter* Vol. 26, Nr. 5, 1970, pp. 217ff.

representative of the physicists said, means

> that [scientific] concepts are not reinterpreted and used in a different sense [from that of the scientists and] that the principles of the Church be made consistent with the findings of the Natural Sciences.

This is Bellarmino's position, except that expert knowledge from a special and rather narrow domain has now taken the place of the wider and more humanitarian point of view of seventeenth-century Catholicism.

According to the second tradition, scientific knowledge is too specialised and connected with too narrow a vision of the world to be taken' over by society without further ado. It must be examined, it must be judged from a wider point of view that includes human concerns and the values flowing therefrom, and its claims to reality must be modified so that they agree with these values. For example: pain, the feelings of friendship, fear, happiness, and the need for salvation, either in secular terms or in terms of some transcendent realm of being, play a large role in human lives. They are basic realities. Hence the claims of some elementary particle physicists to have found the ultimate constituents of everything have to be rejected and replaced by a more 'instrumentalistic' position: their theories are not about reality, they are about predictions in a reality determined independently of their efforts.

At the time of Galileo, this second tradition was the tradition defended by the Church. The Church used its Platonic version: the wider knowledge was expert knowledge but being connected with an eminently human document, the Bible, it had and still has a tremendous advantage over the principles of an abstract rationalism. It is also true that the noble sentiments inherent in a knowledge of this kind did not always prevail and that some Church directives were simply an exercise in power. But the better representatives of the Church thought differently and were worthy predecessors of modern attempts to temper the totalitarian and dehumanising tendencies of modern scientific objectivism by elements directly taken from human life and to that extent 'subjective'.

It must also be admitted — as I have acknowledged — that infringements of epistemological rules are today rarely a matter for the police. However, the law still intrudes, the idea of free and independent research is a chimaera, and the presence or absence of police intervention has nothing to do with the problem before us, namely the interpretation of scientific knowledge-claims. Besides, we have seen (see the brief quotation after footnote 7 above) that even the liberal climate of the modern age has not prevented scientists from demanding the same kind of authority which Bellarmino possessed as a matter of course but exercised with much greater wisdom and grace. It is a pity that the Church of today, frightened by the universal noise made by the scientific wolves, prefers to howl with them instead of trying to teach them some manners.[7]

Finally, a few words about the assertion of some scientists and philosophers that science needs no supervision because it is inherently human and self-correcting. Whatever errors scientists commit, these people tell us, scientists correct; they do so better than any outsider could, and they should therefore be left alone (except of course for a steady influx of the millions they need to pay for their corrections). It is easy to show the flaws in this assertion.

It is of course true that science is inherently human and that a scientist can be as kind or as nasty as the next person. The trouble is that increasing competition inside the scientific establishment, and the increasing attention that is given to the pronouncements of scientists, tend to encourage selfishness, conceit and a contempt for people — for the 'herd', as Galileo called them — who cannot follow the subtle contortions of Nobel minds. The institutionalization of matters previously in the hands of individuals and small groups also encourages opportunism and cowardice. Earlier scientists who were members of religious communities knew that *sub specie aeternitatis* their achievements counted for little. Those modern scientists who combine scientific curiosity with a love for Nature and their fellow human beings to some extent share the sense of perspective of their religious predecessors – but they are surrounded by people with very different ideas. Has science the capability of correcting the

7. In 1982 Christian Thomas and I organised a seminar at the Federal Institute of Technology in Zürich with the purpose of discussing how the rise of the sciences had influenced the major religions and other traditional forms of thought. What surprised us was the fearful restraint with which Catholic and Protestant theologians treated the matter – there was no criticism either of particular scientific achievements or of the scientific ideology as a whole. I commented on this restraint in a letter which is reprinted in the appendix to the present chapter. The Proceedings of the seminar were published under the title *Wissenschaft und Tradition* (Paul Feyerabend and Christian Thomas, eds., Zürich 1983).

aberrations that might arise in a complex enterprise of this kind?

Of course it has – there exists no enterprise that is not self-correcting to some extent and that cannot be changed by the actions of even a few determined individuals. But science is part of larger units; it is part of a city, a region, of entire nations. Depending on their political structure these larger units, too, may be self-correcting. Democracy, especially ancient Greek democracy, was ready to correct everything that occurred in its midst, the effusions of experts included. But democracy, according to the position I am discussing now, has no business interfering with the work of science. Why? One reason given is that this work is much too complex to be understood by laymen. The same can be said about interdisciplinary work within the sciences – and yet such work is encouraged and its results are praised. Many scientists in defending their results use arguments that make a philosopher's hair stand on end – and yet the arguments are accepted and science proceeds on their basis. Besides, institutions such as trial by jury and citizens' initiatives[8] show that laypeople can be instructed, or can instruct themselves, about recondite matters and can in this way acquire the knowledge needed for a balanced judgement. The legalization of acupuncture in California was the result of a learning process of this kind.

A second defence of scientific autonomy is that science is 'objective' and should thus be separated from the 'subjective' opinions of politics (this is an old defence; it can be found in Plato). But a democracy cannot simply bow to the assertions of scientists and philosophers, it must examine these assertions, especially when they touch on fundamental matters; for example, it must examine this claim of 'objectivity'. In other words, it must enter upon a philosophical analysis of scientific claims just as it must enter on a financial analysis of local and national budgets. And in entering upon such an analysis it will have to rely not just on objective truths, but on the way in which these truths present themselves to its members, i.e. it will have to rely on the subjective judgements of these members. To sum up this part of the argument: science, which is self-correcting, is part of larger units which are also self-correcting. In a democracy the self-correction of the larger units includes all their parts, which means that democratic self-correction overrules the temporary

8. Examples of the efficiency of the latter are discussed in Meehan, *The Atom and the Fault*, Cambridge 1984. Common citizens, not experts, encouraged the collaboration between builders and geologists in evaluating the safety of nuclear power plants in California.

results of scientific self-correction.

Self-correction includes a critique of limitless change. The sciences have by now left the qualitative world of our everyday experience far behind. Some scientists claim that this world is a mere appearance and that reality lies elsewhere. They see human beings in terms of this reality and approach them accordingly. But human beings may object to such treatment. They may declare themselves to be a reality different from reality as defined by scientists and may decide to stabilise the former. For example, they may decide to stabilise the qualitative world of our everyday experiences and regard every deviation from it as a step into inhumanity. This is how decisions about the quality of our lives can determine what is to be regarded as real and what is to be regarded as an appearance or a mere instrument of prediction.

The enthusiasm for criticism shown by the philosophers and scientists whose views I am discussing now, though shared by many intellectuals, is not the *only* basis for a rich and rewarding life and it is very doubtful if it can even be *a* basis. Human beings need surroundings that are fairly stable and give meaning to their existence. The restless criticism that allegedly characterises the lives of scientists can be *part* of a fulfilling life, it cannot be its *basis*. (It certainly cannot be a basis of love, or of friendship). Hence, scientists may *contribute* to culture, but they cannot provide its *foundation* — and, being constrained and blinded by their expert prejudices, they certainly cannot be allowed to decide, without control from other citizens, what foundation the citizens should accept. The Churches have many reasons to support such a point of view and to use it for a criticism of particular scientific results as well as of the role of science in our culture. They should overcome their caution (or is it fear?) and revive the balanced and graceful wisdom of Roberto Bellarmino, just as scientists constantly gain strength from the opinions of Democritus, Plato, Aristotle and their own pushy Patron Saint, Galileo.

Appendix

The following is a letter I wrote (in German) to one of the participants of a debate on the modern relation between the sciences and the Catholic Church.

Dear Father Rupert,

I listened with interest to your talk of Thursday last. I was surprised by two features. The one is the speed with which the Church now retreats in the face of scientific results. This phenomenon does not exist within the sciences (although there is much opportunism even here). It often happens that a scientific point of view is shown to be erroneous, but its defenders do not give up, they pursue it for decades, even for centuries, and it often turns out that they are right. The atomic theory is one example. It was frequently 'refuted' but it always returned and defeated its defeaters. Towards the end of the nineteenth century some continental physicists regarded it as a metaphysical monster; the theory was in conflict with facts and internally incoherent. Still its defenders (Boltzmann and Einstein among them) persevered and finally led it to victory. Now if it is legitimate to hold on to and to defend refuted views *within* the sciences, if such a procedure can lead to scientific progress, why, then, does the Church hesitate to do the same thing *from the outside*? For the situation, certainly, is very similar. One of the first scientific attacks against the doctrines of the Church was based on Aristotle's arguments against an origin of the universe. These arguments had much in common with modern cosmological arguments – they rested on known and highly confirmed natural laws and extrapolations therefrom. Long after Aristotle the eternity of the material world as we know it was regarded as a basic fact of science – and then science changed. Today there are numerous world models postulating a beginning in time and leading to complex 'creations' during the first world minutes. The restraint, not to say fearfulness, of the Church, therefore, cannot be excused by pointing to scientific *practice*. It rests on an *ideology* pure and simple. Which brings me to the second point.

You have said that what matters is not so much a certain physics or cosmology but the relation of human beings to God and you have said that this relation has much in common with love. Now, love in the countryside differs from love in the city and there are situations where love becomes almost impossible. For example, love becomes impossible for people who insist on 'objectivity', i.e. who live entirely in accordance with the spirit of science. The sciences encourage objectivity, even demand it; they thereby reduce our ability to love, except in a very intellectual way, which means that a person who wants to disseminate love cannot forget the sciences – (s)he must deal with them and fight certain tendencies inherent in them.

When I was a student I revered the sciences and mocked religion and I felt rather grand doing that. Now that I take a closer look at the matter I am surprised to find how many dignitaries of the Church take seriously the superficial arguments I and my friends once used, and how ready they are to reduce their faith accordingly. In this they treat the sciences as if they, too, formed a Church, only a Church of earlier times and with a more primitive philosophy when one still believed in absolutely certain results. A look at the history of the sciences, however, shows a very different picture.

Best Wishes

Paul Feyerabend

10

Putnam on Incommensurability

1. In his book *Reason, Truth and History* (Cambridge 1981, p.114), Hilary Putnam asserts that 'both of the two most influential philosophies of science of the Twentieth Century . . . are self-refuting'. The philosophies he has in mind are logical positivism and the historical approach. I shall discuss an idea that belongs to the latter, viz. incommensurability, and I shall show that while the idea may have unusual consequences, self-refutation is not one of them.

2. According to Putnam, 'the incommensurability thesis is the thesis that terms used in another culture, say, the term "temperature" as used by seventeenth century scientists, cannot be equated in meaning and reference with any terms or expressions *we* possess' (p.114). I shall call the incommensurability thesis as defined in this statement I.

To refute I, Putnam points out

(A), that 'if [I] were really true, then we could not translate other languages – or even past stages of our language – at all' (p.114), adds

(B), that 'if Feyerabend . . . were right, then members of other cultures, including seventeenth century scientists would be conceptualisable by us only as animals producing responses to stimuli', and concludes

(C): 'To tell us that Galileo has "incommensurable" notions *and then to go on and to describe them at length* is totally incoherent' (p.114f – Putnam's italics).

3. A, B and C rest on the following two assumptions:

[i] understanding foreign concepts (foreign cultures) requires translation and
[ii] a successful translation does not change the translating language.

These two assumptions are characteristic of theoretical traditions and Putnam's arguments, therefore, provide a very good illustration of the general observations made in chapter 3.

Neither [i] nor [ii] is correct. We can learn a language or a culture from scratch, as a child learns them, without detour through our native tongue (linguists, historians and anthropologists, having realised the advantages of such a procedure, now prefer field studies to the reports of bilingual informants). And we can change our native tongue so that it becomes capable of expressing alien notions (successful translations always change the medium in which they occur: the only languages satisfying [ii] are formal languages and the languages of tourists).

Modern lexica exploit both possibilities. Instead of the semantic equations that formed the basis of older dictionaries they employ research articles of an open and speculative nature. (See, for example, the introduction and the major research articles in Bruno Snell et al., *Lexikon des Frühgriechischen Epos,* Göttingen 1971.) Analogies, metaphors, negative characterizations, bits and pieces of cultural history are used to present a new semantic landscape with new concepts and new connections between them. Historians of science proceed in a similar way, but more systematically. Explaining, say, the notion of 'impetus' in sixteenth- and seventeenth-century science, they first teach their readers the physics, metaphysics, technology, and even the theology of the time: in other words, they too introduce a new and initially unfamiliar semantic landscape, and then show where impetus is located in it. Examples are found in the work of Pierre Duhem, Anneliese Maier, Marshall Clagett, Hans Blumenberg and, for other concepts, Ludwik Fleck and Thomas Kuhn.

Translating a language into another language is in many ways like constructing a scientific theory; in both cases we must find concepts that fit the 'language of the phenomena'. In the natural sciences the phenomena are those of inanimate nature. Nobody doubts that it is difficult to give a general account of them, that we may have to revise the terms with which we started and that we may have to revise them further when new phenomena appear. In the case of translation the phenomena are the ideas implicit in another language. These ideas developed in different and often unknown geographical surroundings and under different and often unknown social circumstances, and they went through numerous intended and unintended changes (influence of further languages, deterioration, poetic licence, etc.). Putnam's [ii] assumes that every language contains everything that is needed for dealing with all these eventualities. To use an example, it makes the rather unlikely assumption that modern Swahili is already adapted to the language of the Eskimo and, therefore, to Eskimo history. There are only two ways in which such an assumption could succeed: apriorism, or pre-established harmony. Being an empiricist I reject both.

4. According to Putnam, I makes it impossible to explain foreign (primitive, technical, ancient) concepts in English – this is the content of C. He is right in one sense, wrong in another. It is indeed impossible, and trivially so, to formulate ideas in a language not fit to receive them. But the criteria which identify a natural language do not exclude change. English does not cease to be English when new words are introduced or old words given a new sense. Every philologist, anthropologist, or sociologist who presents an archaic (primitive, exotic, etc.) world view, every popular science writer who wants to explain unusual scientific ideas in ordinary English, every surrealist, dadaist, teller of fairy tales or ghost stories, every science fiction novelist and every translator of the poetry of different ages and nations knows how first to construct, out of English *words*, an English *sounding* model of the pattern of usage he needs and then to adopt the pattern and to 'speak' it. A rather trivial example is Evans-Pritchard's explanation of the Azande word *mbisimo* designating the ability of their poison oracle to see far off things. In his book *Witchcraft, Oracles and Magic Among the Azande* (abridged edition, Oxford 1975, p. 55), Evans-Pritchard 'translates' *mbisimo* as 'soul'. He adds that it is not soul in our sense, implying life and consciousness, but a collection of public or 'objective' events. The addition modifies the use of the word 'soul' and makes it more suitable for expressing what the Azande had in mind. Why 'soul' and not another word? 'Because the notion this word expresses in our own culture is nearer to the Zande notion of *mbisimo* of persons than any other English word' – i.e. because of an *analogy* between the English soul and the Azande *mbisimo*. The analogy is important for it smoothes the transition from the original to the new sense; we feel that despite the change of meaning we are still speaking the same language. Now if a conceptual change like the one just described does not go through a metalanguage but stays in the language itself (in which case we would speak of changing the properties of things rather than the usage of words), and if it is not only a single term but an entire conceptual system that is being received, then we have the situation alluded to in (C), but defused, for the English with which we start is not the English with which we conclude our explanation.

5. Azande ideas already exist in a spoken language and English notions were changed to accomodate them. There are other cases where linguistic change introduces a novel and as yet unexpressed point of view. The history of science contains many examples of this kind. I shall explain the matter by taking an example from the history of ideas.

In *Iliad* 9, 225ff., Odysseus tries to get Achilles back into the battle against the Trojans. Achilles resists. 'Equal fate', he replies, 'befalls the negligent and the valiant fighter; equal honour goes to the worthless and to the virtuous' (318f.). He seems to say that honour and the appearance of honour are two different things.

The archaic notion of honour did not allow for such a distinction. Honour as understood in the epic was an aggregate consisting partly of individual and partly of collective actions and events. Some of the elements of the aggregate were: the position (of the individual possessing or lacking honour) in battle, in the assembly, during internal dissension; his place at public ceremonies; the spoils and gifts he received when a battle was finished and, naturally, his behaviour on all these occasions. Honour was present when (most of) the elements of the aggregate were present, absent otherwise (cf. *Il.* 12, 310ff. – Sarpedon's speech).

Achilles introduces a different point of view. He was offended by Agamemnon, who had taken his gifts. The offence created a conflict between the individual and the collective ingredients of honour. The Greeks who appeal to Achilles, Odysseus among them, illustrate the customary resolution of the conflict: Achilles' gifts have been returned, more gifts have been promised, harmony has returned to the aggregate, honour has been restored (519, 526, 602f.). So far we are squarely within tradition. Achilles moves away from it. Pushed along by his lasting anger, he perceives an equally lasting imbalance between personal worth and social rewards. What he has in mind not only differs from the traditional aggregate, it is not even an aggregate, for there is no set of events that guarantees the presence of honour as he now sees it. Using Putnam's terminology, we can say that Achilles' idea of honour is 'incommensurable' with the traditional idea. And indeed, given the epic background, the short excerpt I quoted from Achilles' speech sounds as non-sensical as the statement 'equal time needs the fast and the slow to reach the goal'. Yet Achilles introduces his idea in the very same language that seems to exclude it. How is that possible?

It is possible because, like Evans-Pritchard, Achilles can change *concepts* while retaining the associated *words*. And he can change concepts without ceasing to speak Greek because concepts are ambiguous, elastic, capable of reinterpretation, extrapolation, restriction; to use a term from the psychology of perception, concepts like percepts obey figure-ground relations.

For example, the tension between the individual and the collective elements of honour that was caused by Agamemnon's deed can be seen in at least two ways — as involving ingredients of equal weight, or as a conflict between fundamental and more peripheral elements. Tradition accepted the first view or, rather, there was no question of a conscious acceptance – people *simply acted* that way: 'With gifts promised go forth. The Achaeans will honour you as they would an immortal!' (602f.). Achilles, driven by his anger, magnifies the tension so that it turns from a transitory disturbance into a cosmic rift (figure-ground relations often change as a result of strong emotions; this is the principle of the Rohrschach test).

The extrapolation does not void his speech of meaning because there exist analogies for what he is trying to express. Divine knowledge and human knowledge, divine power and human power, human intention and human speech (an example used by Achilles himself: 312f.) are opposed to each other as Achilles opposes personal honour and its collective manifestation. Guided by the analogies, Achilles' audience is drawn into the second way of seeing the tension and so discovers, as Achilles did, a new side of honour and of archaic morality. The new side is not as well defined as the archaic notion – it is more a foreboding than a concept – but the foreboding produces new ways of speaking and thus, eventually, clear new concepts (the concepts of some pre-Socratic philosophers are endpoints of this line of development). Forebodings are excluded from theoretical traditions, which therefore either block conceptual change or cannot account for it once it has taken place. Thus if we take the unchanged traditional concepts as a measure of sense, we are forced to say that Achilles speaks nonsense (cf. A. Parry's account in 'The Language of Achilles', *Transactions and Proceedings of the American Philosophical Association*, Vol. 87 (1956) and my own comments in *Against Method*, p. 267). But measures of sense are not rigid and unambiguous and their changes are not so unfamiliar as to prevent the listeners from grasping what Achilles has in mind. Speaking a language or explaining a situation, after all, means both *following* rules and *changing* them; it is an almost inextricable web of logical and rhetorical moves.

From what has just been said, it also follows that speaking a language goes through stages where speaking indeed amounts to merely 'making noises' (Putnam, 122). For Putnam this is a criticism of the views he ascribes to Kuhn and myself (cf. section 2, objection B of the present chapter). For me it is a sign that Putnam, because of his bias in favour of theoretical traditions, is unaware of the many ways in which language can be used. Little children learn a language by attending to noises which, being repeated in suitable surroundings, gradually assume meaning. Commenting on the explanations which his father gave him about questions of logic, Mill writes in his autobiography (quoted from Max Lerner, ed., *Essential Works of John Stuart Mill*, New York 1965, p. 21): 'The explanations did not make the matter clear to me at the time; but they were not therefore useless; they remained as a nucleus for my observations and reflections to crystallize upon; the import of his general remarks being interpreted to me, by the particular instances which came under my notice *afterwards*.' St. Augustine advised parsons to teach the formulae of the faith by rote, adding that their sense would emerge as a result of prolonged use within a rich, eventful and pious life. Theoretical physicists often play around with formulae which do not yet make any sense to them until a lucky combination makes everything fall into place (in the case of the quantum theory we are still waiting for this lucky combination). And Achilles, by his way of talking, created new speech habits which eventually gave rise to new and more abstract conceptions of honour, virtue and being. Thus using words as mere noises has an important function even within the most advanced stages of speaking a language (cf. my *Against Method*, p. 270).

One scientist who was aware of the complex nature of explanatory talk and who used its elements with superb skill was Galileo. Like Achilles, Galileo gave new meanings to old and familiar words; like Achilles he presented his results as parts of a framework that was shared and understood by all (I am now speaking of his change of basic kinematic and dynamical notions); but unlike Achilles he knew what he was doing and he tried to conceal the conceptual changes he needed to guarantee the validity of his arguments. Chapters 6 and 7 of my book *Against Method* contain examples of his art. Taken together with what I have said here, these examples show how it is possible to assert, without becoming incoherent, that the Galilean notions are 'incommensurable' with our own 'and then to go on and to describe them at length'.

6. They also solve Putnam's conundrum about the relation between relativity and classical mechanics. If I is correct, says Putnam, then the sense of statements that occur in a test of either relativity or classical mechanics cannot be 'independent of the choice between Newtonian and Einsteinian theory'. Moreover, it is then impossible to find equations of meaning between 'any word in . . . Newtonian theory [and] any word in . . . general relativi[ty]' (p. 116). He infers that there is no way to compare the two theories.

The inference is again mistaken. As I mentioned in section 3, linguists long ago ceased using equations of meaning to explain new and unfamiliar ideas, while scientists have always emphasized the novelty of their discoveries and of the concepts used in their formulation. This does not stop them from comparing theories, however. Thus the relativist can say that the classical formulae, *properly interpreted* (i.e. interpreted in the relativistic manner), are successful, but not as successful as the full relativistic apparatus. He can argue like a pyschiatrist who, talking to a patient who believes in demons (Newton) adopts his, the patient's, manner of speaking without accepting its demonic (Newtonian) implications (this does not exclude the possibility that the patient will one fine day turn around and convince him of the existence of demons). Or he may teach relativity to the classicist like a foreign language and invite him to judge its virtues from within ('having learned Spanish to perfection and having read Borges and Vargas Llosa, would you not rather write stories in Spanish than in German?'). There exist many other ways in which the Newtonian and the relativist can and do converse. I have explained them in papers written since 1965, some of them in direct response to Putnam's criticisms of that time: see my *Philosophical Papers*, Vol. 1, chapter 6, sections 5ff. and Vol. 2, chapter 8, section 9ff. and appendix. This finishes my response to A, B, and C.

7. The arguments of the preceding sections were based on I, which is Putnam's version of incommensurability. But Putnam's version is not the version I introduced when examining the relation between comprehensive theories, such as Newton's mechanics and relativity or Aristotelian physics and the new mechanics of Galileo and Newton (cf. *Against Method*, pp. 268ff. and *Philosophical Papers*, Vol. 1, chapter 4, section 5). There are two differences. First, incommensurability as understood by me is a rare event. It occurs only when the conditions of meaningfulness for the descriptive terms of one language (theory, point of view) do not permit the use of the descriptive terms of another language (theory, point of view); mere difference of meanings does not yet lead to incommensurability in my sense. Secondly, incommensurable languages (theories, points of view) are not completely disconnected – there exists a subtle and interesting relation between their conditions of meaningfulness. In *Against Method*, I explained this relation in the case of Homeric commonsense versus the language aimed at by the early Greek philosophers. In *Philosophical Papers*, Vol. 1, chapter 4, I explained it in the case of Aristotle and Newton. I should add that incommensurability is a difficulty for philosophers, not for scientists. Philosophers insist on stability of meaning throughout an argument while scientists, being aware that 'speaking a language or explaining a situation means both *following* rules and *changing* them' (see section 5 of the present chapter), are experts in the art of arguing across lines which philosophers regard as insuperable boundaries of discourse.

11

Cultural Pluralism or Brave New Monotony?

In January 1985 I was invited to contribute to a debate about the role of the arts, philosophy and the sciences in the age of postmodernism. In my reply I (a) criticized the assumption that intellectual debates have anything to do with 'world culture', (b) pointed out that the basic phenomenon of 'world culture' is the relentless expansion of Western views and technologies – monotony, not variety is the basic theme of the age, (c) asserted that cultural exchange does not need shared values, a shared language or a shared philosophy, (d) defended variety and 'cacophony' wherever it raised its head and (e) gave a brief account of the development of the philosophy of science from Maxwell to Kuhn. There is no need to reproduce my reply as all these matters are dealt with elsewhere in the present book. But I think that the following letter, which I wrote in response to a long criticism of my reply, raises new issues and may be of some interest.

3.8.85

Dear Messrs. Vergani, Shinoda and Kesler,

Thank you for your long and detailed letter and for the care you have taken to respond to my little pamphlet. Naturally, I don't agree. Let me explain why.

You ask: 'But can you actually deny the importance of a

coherent structure organizing diversity?' I reply that this is neither for you nor for me to decide – it is up to the people who have created the diversity and who are now living in its midst. If the nations on the African continent are content to live side by side, without any cultural or other contact, then this is what they will do no matter how distant 'thinkers' may view their conduct. If Americans like the 'inflation of goods, images, ideas, traditions', look forward to new and improved hairsprays, car models and soap operas, and use money as the ultimate measure of value, then a dissenting intellectual may of course plead with them like a preacher but he becomes a tyrant if he tries to use more powerful means of persuasion. 'We, the philosophers,' wrote Edmund Husserl in a remarkable essay with the title *The Crisis of the European Sciences and Transcendental Phenomenology* (1936), 'are *functionaries of humanity* [his emphasis]. The quite personal responsibility of our own true being as philosophers, our inner personal vocation, bears within itself at the same time the responsibility for the true being of humanity.' You may agree with this quotation. I think it shows an astounding ignorance (what does Husserl know of the 'true being' of the Nuer?), a phenomenal conceit (is there any single individual who has sufficient knowledge of all races, cultures, civilizations to be able to speak of 'the true being of humanity'?) and, of course, a sizeable contempt for anybody who lives and thinks along different lines.

It is true that nations and groups within a society frequently establish some kind of contact, but it is not true that in doing this they create, or assume, a 'common metadiscourse' or a common cultural bond. The connections may be temporary, *ad hoc*, and quite superficial: White South African leaders, Black Muslims and European terrorists all have a great fondness for the dollar – but there is little else that unites them.

Even a closer connection between cultures A,B,C, etc. need not be 'organized' by a 'coherent structure'; all that is needed is that A interact with B, B with C, C with D and so on, where the mode of interaction may change from one pair to the next and even from one episode of interaction to the next. The use of Accadian during the First Internationalism is a case in point. It was not a necessary presupposition of this civilization, it was one of its many features; it happened to be used by special groups who wrote down what they did and therefore caught the eye of scribblers working millenia after them, namely our own scholars. Not every exchange was in Accadian: there were local contacts using the dialects or languages of a narrow region, and even they extended only as far as the needs and the curiosity of the participants. Besides, one must not confound a culture with its written manifestations, or with the products of its artists and its thinkers: the written laws of Hammurabi had little influence on legal practice; the first uses of Linear B were purely commercial, but Greek education was not based on business ideas, it was based on Homer, i.e. on oral poetry. (Some later poets, Plato among them, never got entirely reconciled to having their work distributed in written form – cf. Plato, *Phaedrus* 274d ff., and the Seventh Letter, esp. 241b ff.). Today we have lots of Marxist books and lots of Marxist ideas floating around in our universities and research centres – but does this taint 'our culture' with Marxism? I don't think so, for there is not a trace of Marxism in our soap operas and on our religious channels. *Intellectuals do not yet a culture make.* Of course, we may stipulate that culture equals literature plus art plus science – but then we decide the question of the cultural effects of literature and so on by fiat, not by research.

I also admit that contacts occasionally became much stronger and led to cultural unities in the sense you seem to have in mind – but watch carefully how this was achieved: most of the time the unity was imposed by force, it only rarely emerged from the wishes and the actions of the people concerned. Scientists, artists, run-of-the-mill intellectuals do not seem to object to such developments, they may even encourage them – this is why they try to infiltrate government agencies, this is why they get so upset when their products are subjected to public control, this is why they admire cultural 'Leaders' who share their ideology and their wish for power, and support them with power plays of their own. Your yearning for a new 'metadiscourse' smells dangerously like a rehash of the machinations of Constantine the Great or of the 'education' of the American Indians. I, on the other hand, prefer a form of life where unities arise from the fortuitious merging of temporary links and where they may decay the moment the links are no longer popular.

My next point is that you seem to be unable to make up your mind about the present condition of 'world culture'. In your *Editorial Statement* you say that there is 'cultural chaos' without a unifying bond. But in your letter you insinuate that there might be some bond, only it is a bond you despise (money). I agree with your letter: there is increasing uniformity, not only in the so-called 'first' world (a nice conceit, that, calling a pushy latecomer the 'first' world!) but elsewhere as well, and all the differences and pluralisms which also exist vanish by comparison. They are amusing little confusions, they will hardly disturb General Motors, or Proctor and Gamble, or the Pentagon. Yet even in your letter you repeat that chaos 'pervades the essence of our culture'. That sounds nicely abstract and philosophical but I wonder how closely you have considered the matter. Have you compared the clientèle of soap operas, or of Reverend Falwell, or of the Super Bowl with the clientèle of modern art, or of the rationalism/irrationalism issue in philosophy? Do you have the numbers? The strength of commitment? The influence on the lives of the rest of us?

I don't think you have. Neither have I, but even the simplest calculation shows that you cannot possibly be right: there are now about 10,000 philosophers teaching in the US and in Canada. Most of them are obedient servants of the status quo, but let us assume that 25% are creators of disorder – a vastly exaggerated estimate. Let us also assume that each chaoticist has 100 students. Most of these students will take philosophy because they have to, they will be bored to death and will be glad when the classes and the examinations are over – but let us again assume that 25% become committed followers of their chaos-creating teachers. That would make 40,000. Do you know how many millions are watching *Dallas*? How many watched the Super Bowl? Do you know the number of followers of all TV preachers taken together? Do you remember how many people voted for Reagan? How many people are still supporting his policies? The order of magnitude here is tens of millions – incomparably greater than the already quite exaggerated number I have arrived at in my small calculation. Or compare the amount of money that is being used to uphold chaos with the amount of money that supports monotony. The percentages for Defence and for contributions to the arts of the Gross National Product would give a first approximation; they show how little the arts and the humanities amount to – and the forces of chaos are only a minuscule part of them. Don't tell me that numbers don't count; artists and researchers in the basic sciences constantly harp on numbers when trying to show how little attention they get. I use the same argument to contest your thesis of the pervasiveness of chaos.

(Incidentally, you should not be so modest and explain your conflation of 'world culture' and 'first world culture' by 'editorial inexperience' – scholars with numerous honours on their heads and tons of books and articles to their credit have talked and are still talking in exactly the same way. Just read again the short quotation from Husserl which I gave above. These people speak of 'culture', or of 'Man' – but what they mean are they themselves and those few select creatures who can understand their papers: so, you see, you are in excellent company.)

I now come to my last disagreement with you. You 'believe in the autonomy of art, thought, and feeling over money'. Again you write in a manner that sounds impressive but gives no hint of what the implications for the real world are supposed to be. In the real world an artist needs money: money for the rent, for food, colours, brushes, visits to museums, perhaps he or she has to support a mistress or a wife, a lover or a husband, occasionally even both, he or she may have children – and so on. The same is true of philosophers, dancers, movie directors, script writers, poets. All these people need *and want* bigger salaries and/or better prices for their products.

So what do you mean by autonomy? Do you mean that an artist should do without money and starve, or live in a rathole? Do you mean that he or she should be fed and housed but without money being involved? On the basis of barter, for example? Well, that depends on the artist. If he likes to live in a shack in the countryside and keep a cow for milk, then more power to him but, alas, this again needs money to start with. Do you mean that it would be better if there were no money around? That is an interesting dream but of no relevance to our problem, for the question is not how artists might live in a never-never-land, but how they can live here and now, in 1985, in this country. And here and now money is of the essence. Money is not inherently bad; it is a means to an end. It has been put to bad uses and some people are so fascinated by it that they devote their whole lives to its accumulation. I assume that our artist is not one of these people (he would not be a worse artist if he were – Giotto quarreled a lot about money and took great care to increase his riches, and yet he was one of the greatest artists who ever lived).

So, our artist will use money, but he will not revere it like a God. Who will pay him? Perhaps a rich sponsor. In this case the artist may have to adapt his art to the wishes of his patron. Now when speaking of 'autonomy' do you mean to say that the patron has no right to make his wishes known? Because the artist by the very fact of being an artist is above judgement by other people? That is sheer elitism, I reject it and I reject the contempt for others it implies. I reject elitism even more emphatically where public money is involved: a person who is paid from public funds must be ready to accept public supervision. I have the uneasy feeling that when speaking in your abstract manner about 'autonomy', miles removed, apparently, from anything so low and filthy as money, you actually want the public to pay artists (and scientists, and great 'feelers') to live and work as they, the artists etc., see fit, i.e. *as parasites*: academics, using the magic hood of 'academic freedom', have succeeded, long ago, in making parasitism respectable – now the artists want a piece of the action. I am against parasitism (unless agreed upon by all parties) and this means that I am against academic freedom and, naturally, against any corresponding 'artistic autonomy'.

'But great art', you may say, 'assumes the complete autonomy of the artist'. Not so, as is shown by Renaissance artists who had to obey city fathers and private money-bags and by composers such as Haydn or Mozart who wrote *Gebrauchsmusik*, were paid for it and yet produced some of the greatest art of your beloved 'first world'. 'But today', you might continue, 'the situation is different. Today the public has no taste – witness the popularity of *Dallas* and *Dynasty*.' Here you argue from Contempt. *Dallas* and *Dynasty* are the art of the masses – true. But the masses consist of individuals, so you can either say 'individuals like you and me' – then you are a humanitarian and will respect their choices – or you can say 'individuals without taste' – then you are conceited bastards and why, then, should the masses pay you? Besides, a good movie, that is, a movie that is artistic without pandering to the recondite tastes of a few sublime souls – and there are lots of such movies – shows how close the collaboration between great art and good money can be. The collaboration is not easy – few important things are – but it can be fruitful and it was such a fruitful collaboration and not foggy (but at bottom contemptuous) demands for autonomy that gave us the arts of the past.

One more point before I take my leave. I frequently watch the discussions and the audience reactions at the Phil Donahue show. These are ordinary people, they watch TV, they go to the movies, many of them support the one or the other policy of Ronald Reagan, many are religious people, they work hard to earn an income, to bring up their children, to support their relatives. I also read authors such as Russell Baker (his autobiography) and Evelyn Keyes (her autobiography). They, too, speak about human affairs, they speak clearly, simply and in concrete terms; they have heart, they show wisdom, understanding, they are often puzzled, they do not know, and they say so, they do not hide their puzzlement behind empty words. Now the interests of all these people and your own interests seem to be remarkably similar – you all are concerned about adverse developments – but what a difference in language! Simple and personal narration on the one side, an uneasy mishmash of impersonal abstractions on the other. I know what experts are saying about this contrast. They say that social analysis is a difficult matter and that it needs a severely theoretical discourse to succeed. I reply that a theoretical discourse makes sense in the natural sciences where abstract terms are summaries of readily available results but that theoretical statements about social affairs often lack content and become either nonsensical or trivially false when the content is provided (cf. my brief comments on your main thesis and on your plea for artistic autonomy). The wall of incomprehension erected by such talk is therefore not based on knowledge, it is based on pretence and on the wish to intimidate – one more reason to take a very critical look at the many privileges intellectuals have managed to steal in our society.

And now all the best to you and all your further enterprises!

12

Farewell to Reason

The German version of this essay was based on the third German edition of Against Method *(AM for short) which differs from the English, the French, the Japanese and the Portuguese editions and was published in 1986.* Erkenntnis Für freie Menschen *(EFM for short) is the largely (two-thirds) rewritten German edition of* Science in a Free Society *(SFS). It does not contain the chapters on Kuhn, Aristotle and Copernicus and the replies to critics which constituted more than half of the English text. Instead there is a more detailed explanation of the relation between reason and practice, an extended chapter on relativism and a sketch of the rise of rationalism in antiquity. The criticisms on which I comment were published in H. P. Doerr (ed.)* Versuchungen, *2 vols, Frankfurt 1980/81.*

1 Survey

This chapter deals with the following topics: the structure of scientific reasoning and the role of a philosophy of science; the authority of science compared with other forms of life; the importance of such other forms of life; the role of abstract thought (philosophy, religion, metaphysics) and abstract ideals (humanitarianism, for example). It also contains replies to critical essays that appeared in German in 1980 and clarifies points made in AM and EFM.

2   The Structure of Science

My main thesis on this point is: the events and results that constitute the sciences have no common structure; there are no elements that occur in every scientific investigation but are missing elsewhere (the objection that without such elements the word 'science' has no meaning assumes a theory of meaning that has been criticized, with excellent arguments, by Ockham, Berkeley and Wittgenstein).

Concrete developments (such as the overthrow of steady state cosmologies or the discovery of the structure of DNA) have of course quite distinct features and we can often explain why and how these features led to success. But not every discovery can be accounted for in the same manner and procedures that paid off in the past may create havoc when imposed on the future. Successful research does not obey general standards; it relies now on one trick, now on another, and the moves that advance it are not always known to the movers. A theory of science that devises standards and structural elements of *all* scientific activities and authorizes them by reference to some rationality-theory may impress outsiders – but it is much too crude an instrument for the people on the spot, that is, for scientists facing some concrete research problem. The most we can do for them from afar is to enumerate rules of thumb, give historical examples, present case studies containing diverging procedures, demonstrate the inherent complexity of research and so prepare them for the morass they are about to enter. Listening to our tale, scientists will get a feeling for the richness of the historical process they want to transform, they will be encouraged to leave behind childish things such as logical rules and epistemological principles and to start thinking in more complex ways – and this is all we can do *because of the nature of the material*. A 'theory' of knowledge that intends to do more loses touch with reality. Not only *are* its rules not used by scientists, they *cannot* possibly *be* used in all circumstances — just as it is impossible to climb mount Everest using the steps of classical ballet.

The ideas just presented (and illustrated with historical examples in AM and in my *Philosophical Papers*, Cambridge 1981) are not new. As I wrote in section 4 of chapter 6, we find them in philosophers like Mill (his *On Liberty* – the outstanding presentation of a libertarian epistemology), in scientists such as Boltzmann, Mach, Duhem, Einstein and Bohr, and then, in a philosophically already quite desiccated way, in Wittgenstein. They were fruitful ideas: the revolutions of modern physics, relativity and quantum mechanics and the later changes in psychology, biology, biochemistry and high energy physics would have been impossible without them. Yet they had only a slight impact on philosophy. Even the most inconoclastic philosophical movement of the time, neopositivisim, still clung to the ancient idea that philosophy must provide general standards for knowledge and action and that science and politics can only profit from adopting such standards. Surrounded by revolutionary discoveries in the sciences, interesting points of view in the arts and unforeseen developments in politics, the stern fathers of the Vienna Circle withdrew to a narrow and badly constructed bastion. The connection with history was dissolved; the close collaboration between scientific thought and philosophical speculation came to an end; terminology alien to the sciences and problems without scientific relevance took over.

Fleck, Polanyi and then Kuhn were (after a long time) the first thinkers to compare the resulting school philosophy with its alleged object – science – and to show its illusionary character. This did not improve matters. Philosophers did not return to history. They did not abandon the logical charades that are their trademark. They enriched these charades by further empty gestures, most of them taken from Kuhn ('paradigm', 'crisis', 'revolution', and so on) without regard for context, and thus complicated their doctrine; but they did not bring it closer to reality. Pre-Kuhnian positivism was infantile, but relatively clear (this includes Popper who is just a tiny puff of hot air in the positivistic teacup). Post-Kuhnian positivism has remained infantile – but it is also very unclear.

Imre Lakatos was the only philosopher of science to take up Kuhn's challenge. He fought Kuhn on his own ground and with his own weapons. He admitted that positivism (verificationism, falsificationism) neither enlightens scientists nor aids them in their research. However, he denied that stepping closer to history forces us to relativize all standards. This may be the reaction of a confused rationalist who for the first time faces history in its full splendour but, so Lakatos said, a more thorough study of the same material shows that scientific processes share a structure and obey general rules. We can have a theory of science and, more generally, a theory of rationality because thought enters history in a lawful way.

In AM as well as in chapter 10 of vol. 2 of my *Philosophical Papers* I tried to refute that thesis. My procedure was partly abstract, consisting in a criticism of Lakatos' interpretation of history, and partly historical. Some critics deny that the historical examples support my case: their objections will be dealt with below. However, if I am correct – and I am pretty sure I am – then it is necessary to return to the position of Mach, Einstein and Bohr. A theory of science is then impossible. All we have is the process of research and, side by side with it, all sorts of rules of thumb which *may* aid us in our attempt to further the process but which may also lead us astray. (What are the criteria that inform us that we have been misguided? They are criteria which seem to fit the situation at hand. How do we determine fitness? We *constitute* it by the research we do: criteria do not merely *judge* events and processes, they are often constituted by them and they must be introduced in this manner or else research could never get started: AM, 26.)

This is my simple answer to various critics who either chastise me for opposing theories of science and yet developing a theory myself, or take me to task for failing to give a 'positive determination of what good science consists in': if a collection of rules of thumb is called a 'theory', well, then of course I have a 'theory' – but it differs considerably from the antiseptic dream castles of Kant and Hegel and from Carnap's and Popper's dog huts. Mach, Einstein and Wittgenstein, on the other hand, lack a more impressive edifice of thought not because they are lacking in speculative power, but because they have realized that freezing this power into a system would mean the end of the sciences (the arts, religion, and so on). And the natural sciences, especially physics and astronomy, enter the argument not because I am 'fascinated by them', as some confused champions of the humanities remarked, but because they are the issue: they were the weapons which the positivists and their anxious foes, the 'critical' rationalists, trained on unloved philosophies, and they are the weapons which now cause their own demise instead. Nor do I speak of progress because I believe in it or pretend to know what it means (using a *reductio ad absurdum* does not commit the arguer to accepting the premises: cf. AM, p. 27). As for the slogan 'anything goes', which certain critics have attributed to me and then attacked: the slogan is not mine and it was not meant to summarise the case studies of AM and SFS. I am not looking for new theories of science, I am asking if the search for such theories is a reasonable undertaking and I conclude that it is not: the knowledge we need to understand and to advance the sciences does not come from theories, it comes from participation. The examples, accordingly, are not details that can and should be omitted once the 'real account' is given – they *are* the real account. The critics, holding a belief I explicitly reject (that there can be a theory of science and of knowledge), read only part of my story and they read it in a way that is contradicted by the rest. Small wonder they are baffled by the result.

Similar remarks apply to readers who accept the slogan and interpret it as making research easier and success more accessible. My objection to these lazy 'anarchists' is again that they misread my intentions: 'anything goes' is not a 'principle' I defend, it is a 'principle' forced upon a rationalist who loves principles but who also takes history seriously. Besides, and more importantly, an absence of 'objective' standards does not mean less work; it means that scientists have to check *all* ingredients of their trade and not only those which philosophers and establishment scientists regard as characteristically scientific. Scientists can thus no longer say: we already have the correct methods and standards of research – all we need to do is to apply them. For according to the view of science that was defended by Mach, Boltzmann, Einstein and Bohr, and which I restated in AM, scientists are not only responsible for the correct *application* of standards they have imported from elsewhere, they are responsible *for the standards themselves*. Not even the laws of logic are exempt from their scrutiny for circumstances may force them to change logic as well (some such circumstances have arisen in the quantum theory).

This situation must be kept in mind when considering the relation between 'great thinkers' on the one hand and editors, moneybags and scientific institutions on the other. According to the traditional account, scientists with uncommon ideas and the institutions from which they seek support have certain general ideas in common: they are both 'rational'. All a scientist in search of money has to do is to show that his research, apart from containing novel suggestions, conforms to these ideas. According to the account defended by me, the scientists and their judges must first establish some common ground – they can no longer rely on standard slogans (their exchange is 'free', not 'guided': cf. SFS, p. 29).

In such a situation the demand of 'anarchic' scientists for

greater freedom can be interpreted in two ways. It can be interpreted as a demand for an open exchange that seeks understanding without being tied to specific rules. But it can also be interpreted as a demand for acceptance without examination. In the terms of AM and SFS, the latter demand might even be supported by pointing out that ideas that were once regarded as absurd have subsequently led to progress. The argument overlooks that the judges, editors or moneybags can use the same reasons: the status quo, too, has led to progress and 'anything goes' includes the methods of its defenders. It is therefore necessary to offer a little more than arrogance and vague generalities.

The case studies show that scientific rebels took this extra step. Galileo, for example, did not just complain, he tried to convince his opponents with the best means at his disposal. These means frequently differed from standard professional procedures, they even conflicted with commonsense – here is the anarchic component of Galileo's research; but they had a reason of their own which could be expressed in commonsense terms and they were occasionally successful. And let us not forget that a full democratization of science will make life even more difficult for the self-proclaimed discoverers of Great Ideas, who will then have to address people who do not even share their interest in science or research. What will our freedom-loving 'anarchists' do under such circumstances? When their opponents are no longer hated big-shots but much beloved free citizens?

3   Case Studies

In this section I shall deal mainly with objections to my treatment of Galileo. Let me repeat that I criticize not Galileo's procedures – which are excellent examples of the inventiveness of scientific practice mentioned in section 2 – but those philosophical theories which, if applied with a better knowledge of history, would have to reject them as 'irrational'. Galileo was irrational in terms of these theories – but he was also one of the greatest scientist-philosophers who ever lived.

According to Gunnar Andersson, the Galileo case may endanger an 'overly simple and naive version of falsificationism' – but it does not threaten a philosophy where both theories and observations are fallible. My interpretation of Galileo's assumptions further reveals, according to Andersson, that I have not understood Popper's definition of *ad hoc* hypotheses. *Ad hoc* hypotheses, says Andersson, are not merely introduced to explain special effects; they also lower the degree of falsification of the system in which they occur.

Now that is precisely what Galileo's assumptions do. Galileo's account of motion turns the tower argument[1] from a refutation of Copernicus into a confirming instance and reduces the content of the Aristotelian dynamics that preceded it (AM, pp. 99f). This latter theory (explained in books i, ii, vi, and viii of the *Physics*) deals in a general way with a variety of changes including locomotion, generation, corruption, qualitative change (such as the transmission of knowledge from a well informed teacher to an ignorant pupil – an example often used by Aristotle), increase and decrease. It contains theorems such as: every motion is preceded by another motion; there exists a hierarchy of motions which starts from an unmoved cause of motion, is followed by a primary motion with constant (angular) velocity and branches out from there; the length of a moving object has no precise value – ascribing to an object a precise length means assuming that it is at rest; and so on. The first theorem was proved by assuming that the world is a lawful entity. (The proof can be used today against the Big Bang theory of the origin of the universe or against Wigner's idea that the reduction of the wave packet is due to an act of consciousness.) The last theorem which is based on Aristotle's account of continuity anticipates basic ideas of the quantum theory (cf. chapter 8 for details).

Aristotle's theory of motion is coherent and it was confirmed to a high degree. It stimulated research in physics (electricity – cf. J. L. Heilbron, *Electricity in the 17th and 18th Centuries*, University of California Press 1979), physiology, biology and epidemiology down to the late nineteenth century and it remains relevant today: the mechanical views of the seventeenth and eighteenth centuries and their modern sequels are incapable of dealing even with their own prize process, locomotion (cf. the work of Bohm and Prigogine as well as chapter 8 of the present

1. According to the Tower Argument (AM, chapter 7) a stone dropped from a tower on a moving earth will be left behind. It is not left behind, hence the earth does not move. The argument assumes (Aristotle's law of inertia) that an object outside the reach of forces remains at (returns to) rest. At the time of the debate this assumption was confirmed. It was used for a considerable time after the Copernican Revolution to establish the existence of flies' eggs, bacteria, viruses.

book). What did Galileo do? He replaced this complex and sophisticated theory, which already contained the distinction between laws of inertia (they describe what happens when no forces are acting) and laws of forces (they describe how forces influence motion), by his own law of inertia which lacked corroboration, applied to locomotion only and 'drastically reduced the degree of falsification of the entire system.'

Concerning the falsifiability of observational statements the situation is, however, as follows. Critical rationalism, the 'philosophy' defended by Andersson, is either a fruitful point of view that guides scientists, or it is empty talk that can be reconciled with any procedure. Popperians say it is the first (rejection of Neurath's assertion that any statement can be crossed out for any reason whatsoever). This is why they insist that basic statements intended to refute a theory must be highly corroborated. Galileo's telescopic observations did not satisfy this demand: they were self-contradictory, not everybody could repeat them, those who did repeat them (Kepler) got puzzling results and there existed no theory to separate 'phantoms' from veridical phenomena (physical optics, mentioned by Andersson, is irrelevant; the basic statements under discussion are not about rays of light but about the position, colour and structure of visual patches, and a popular hypothesis correlating the first with the second could be easily shown to be false: cf. AM, p. 137). Galileo's basic statements are therefore bold hypotheses without much corroboration. Andersson accepts this description – it needs time, he says, to obtain the corroborating evidence (and the related touchstone theories, to use an excellent expression of Lakatos's). The first interpretation of critical rationalism mentioned above asserts that during the search the statements have no refuting power. If one still says, as Andersson does, that Galileo refuted popular views by his observations, then one moves from the first interpretation to the second, where basic statements can be used in any way whatsoever. The verbiage remains critical – but its content has evaporated.

Next comes a criticism which T.A. Whitaker has published in two letters in the journal *Science* (May 2 and October 10, 1980). Whitaker points out that there exist two sets of pictures of the moon, the woodcuts (which I presented in AM) and the copperplates which are more accurate, from a modern point of view. The copperplates, says Whitaker, show Galileo to be a better observer of the moon than I make him out to be.

Now, first of all, I never doubted Galileo's ability as an

observer. Quoting R. Wolf, who writes that 'Galileo was not a great astronomical observer; or else the excitement of the many telescopic observations made by him at the time had temporarily blurred his skill or his critical sense', I reply (AM, p. 129) that

> this assertion may well be true (though I rather doubt it in view of the extraordinary observational skill which Galileo exhibits on other occasions). But it is poor in content and, I submit, not very interesting . . . There are however other hypotheses which lead to new suggestions and which show us how complex the situation was at the time of Galileo.

I then mention two such hypotheses, one dealing with general features of contemporary telescopic *vision*, the other considering the assumption that perceptions, i.e. the things seen with the naked eye, have a *history* (which may be discovered by combining the history of visual astronomy with the history of painting, poetry, etc.).

Secondly, reference to the copperplates does not remove all the troublesome aspects of Galileo's observations (of the moon). Galileo not only drew pictures, he also gave verbal descriptions. For example, he asked (cf. AM, p. 127): 'Why don't we see unevenness, roughness and waviness in the waxing moon's outmost periphery which faces West, in the waning moon's other circular edge which faces East and in the full moon's entire circumference? Why do they appear perfectly round and circular?' Kepler replied, on the basis of naked-eye observations (cf. AM, p. 127 fn. 24): 'If you look carefully at the moon when it is full, it seems perceptibly to be lacking in roundness', and he answered Galileo's question by saying: 'I do not know how carefully you have thought about this subject or whether your query, as is more likely, is based on popular impression. For . . . I stated that there was surely some imperfection in this outermost circle during full moon. Study the matter again and tell us how it looks to you.'

This little exchange shows, thirdly, that the problems of observation which existed at Galileo's time cannot be solved by showing that Galileo's observations agree with *our* view of the matter. To see how Galileo proceeded, if he was 'rational' or if he broke important rules of scientific method, we have to compare his achievements and his suggestions with *his* surroundings and *not* with the situation in an as yet unknown future. If it turns out that the phenomena reported by Galileo were not confirmed by anyone else, and that there were no reasons for trusting the telescope as an instrument of research but many reasons, both theoretical and observational, that spoke against it, then it was as unscientific for Galileo to push the phenomena as it would be unscientific for us to push experimental results which lack independent corroboration and are obtained by doubtful methods – no matter how closely his observations approach our own. For to be scientific in the sense that is at issue here (and that *is criticized* in AM and SFS) means to act properly with respect to existing and not with respect to possible knowledge.

Now I used the woodcuts in order to gauge the reactions of Galileo's contemporaries. Note again that I did not try to argue that Galileo was a lousy scientist because the woodcuts differ from modern pictures of the moon – such an argument would have conflicted with the considerations just given. My assumption was, rather, that the moon as seen with the naked eye looks different from the woodcuts, that it might have looked different to Galileo's contemporaries and that some of them might have criticised the *Sidereus Nuncius* on the basis of their own naked-eye observations. This assumption is still useful, for the woodcuts accompanied most editions of the book. Does it apply to the engravings as well? It does, as is shown by Kepler's criticism.

There were, moreover, many reasons why the telescope was not uniformly regarded as a reliable producer of facts (some of these reasons, both empirical and theoretical, are assembled in AM). Whitaker's assertion, made in his second communication, that Galileo's drawings of the moon are of a high quality when compared with modern pictures has no bearing on this discussion.

John Worral ascribes to me the 'truism that "theoretical facts" are dependent on theory' as well as arguments that 'depend on taking "fact" at a very high theoretical level'. What I actually assert in the paper in which these matters are explained (now reprinted as chapter 2 of vol. 1 of my *Philosophical Papers*) is that *all* facts are *theoretical* (or, in the formal mode of speech, 'logically speaking all terms are "theoretical" ' – op. cit., p. 32, fn. 22) and not merely theory-*laden*. I also argue for this assertion and show that and why it is preferable to alternatives, including the alternative Worral seems to have in mind. Worral's complaints nowhere touch this position and these arguments.

John Worral's difficulties show how little Popperians have advanced beyond more naive forms of empiricism. Worral wants to distinguish empirical facts and theoretical facts but he does not know how to proceed. Occasionally he proceeds psychologically, i.e. he distinguishes between facts that are accepted by all experts in a certain domain and more doubtful facts which give rise to debates. This Carnap (in *Testability and Meaning*) and I (in section 2 of the abovementioned paper) had done before him, and in a much clearer way. On other occasions he seems to assume that agreements reached go beyond psychology and are grounded in the facts themselves: empirical facts are less pervaded by theory than theoretical facts, they have an 'empirical core'. Neurath, Carnap and I would say that such facts *appear* to be less pervaded with theory: the old Greeks perceived their gods directly – the phenomena did not contain any theoretical element – but philologists eventually discovered the complex ideology at their basis and showed how even very simple divine 'facts' were constituted by a highly complex structure (AM, Ch. 17). Classical physicists described and we still describe our surroundings in a language which neglects the relation between observer and observed object (we assume stable and unchanging things, we base our experiments upon them) but the theory of relativity and the quantum theory showed that this language, this mode of perception and this manner of carrying out experiments rested on cosmological assumptions. The assumptions were not explicitly formulated – this is why we don't notice them and simply speak of empirical 'facts' – but they underlie all phenomena: the apparently empirical 'facts' are theoretical through and through. Yet they frequently function as judges between alternative views.

Worral assumes that judges must be neutral (hence the need for a solid empirical 'core') – i.e. he assumes that scientists who use facts when examining a variety of theories do not change them in the course of the examination. The assumption can easily be shown to be wrong. Relativists and aether theoreticians have different facts, even in the domain of observation. For the relativist, observed masses, lengths, or time intervals are projections of four-dimensional structures into certain reference systems (cf. Synge in de Witt and de Witt, *Relativity, Groups and Topology*, New York 1964), while the 'absolutists' regard them as inherent properties of physical objects. Relativists admit that classical *descriptions* (which were designed to express classical facts) may occasionally be used to convey information about relativistic facts and they employ them in the relevant circumstances. But this does not mean that they accept their classical *interpretation*.

On the contrary, their attitude is very close to that of a phychia-trist who talks to a patient claiming to be possessed in his, the patient's, language, without accepting an ontology of devils, angels, demons and so on: our ordinary ways of speaking, scien-tific arguments included, are much more elastic than is imagined by Worral.

The tower argument, according to Worral, was defused by Galileo in the following way: the moving earth taken in con-junction with the Aristotelian theory of motion (according to which an object not under the influence of forces comes to rest) increases the distance between the stone and the tower. The stone does not move away from the tower. Therefore, says Worral's Galileo, 'the experiment does not refute Copernicus, but a more complex theoretical system' and he replaces Aris-totle's dynamics, which is part of this system, by his own law of inertia. Here he remains within the framework of Duhem's analysis of theory-change. More especially, he corrects a 'logical error' of the anti-Copernicans according to whom the false state-ment (the stone moves away from the tower) follows directly from the assumption that the earth rotates. So far John Worral.

Now first of all, the alleged 'logical error' was never committed by the anti-Copernicans. Being good Aristotelian logicians they knew very well that the derivation needed at least two premises. They even mentioned them explicitly, but they directed the arrow of falsification against only one premise – the motion of the earth – as the other was theoretically plausible and confirmed to a high degree and, besides, it was not the topic at issue (cf. Popper's comments on Duhem's argument against simple falsifi-ability). Secondly, the replacement of Aristotle's law of inertia was only part of the change carried out by Galileo. The Aris-totelian law described absolute motions – and so did the tower argument (the predicted deviation of the stone from the tower is of course a relative change, but the problem under debate here is what Galileo changed and not what reasons he used when carry-ing out the change). If a new 'auxiliary hypothesis' is introduced then this hypothesis, too, must use absolute motions: it must be a form of the impetus theory. But Galileo gradually became a kinematic relativist (AM, p. 78 fn. 10; p. 96, fn. 15). *His* auxiliary hypothesis had to work *without* impetus. Thus he did not merely change one *hypothesis,* of an otherwise unchanged conceptual system (absolute motion around the earth, or around the sun, but not straight towards the centre); he also replaced the *concepts* of the system – he introduced a new world view (which had been prepared by others). The first process can be explained by Duhem's scheme, the second cannot.

Worral also criticizes the way in which I use Brownian motion to argue for a plurality of theories. This criticism is such a wonderful example of the shortcomings of a purely philosophical approach (as described by me in Vol. 2, chapter 5 of my *Philosophical Papers*) that it deserves our fullest attention.

In chapter 3 of AM, I showed that Brownian motion contradicts the second law of phenomenological thermodynamics only when analysed by the kinetic theory which also contradicts that law. Worral says he does not understand my argument. So far so good. There are many things which many people do not understand. In order to understand the argument, Worral translates it into a language he is familiar with, a kind of pidgin logic. This, too, is unobjectionable: if I don't understand an argument, then I shall try to reformulate it in my own way. Worral goes further. He complains that I did not formulate my argument in his language in the first place. But my argument was not part of a personal letter to him, it was addressed to physicists favouring a theoretical monism – and they seem to understand it perfectly. Besides, Worral does not just object to having been left out, he assumes that the language he understands is the only reasonable language there is. In this he is certainly mistaken, as is shown by the nonsense his translation produces (his notion of evidence, for example, makes it impossible to speak of unknown evidence or of events which, although well known, and although evidence, are not known to be evidence). Like a native speaker of a language too poor to express certain states of affairs, he projects the lacuna on to my argument and claims to have shown its incoherence. I, on the other hand, would conclude that there are better languages than pidgin logic. Using one such language my argument can be stated as follows.

Assume we have a theory T (and by this I mean the entire complex: theory plus initial conditions plus auxiliary hypotheses and so on). T says that C will occur. C does not occur, C' occurs instead. If this fact were known then one could say that T was refuted and C' would be the refuting evidence (note that I do not distinguish between facts and statements; no step in the argument depends on the distinction and no intelligent person will be confused by its absence). Assume further that there are laws of nature which prevent us from directly separating C and C': there exists no experiment that could tell us the difference. Finally, let us assume that it is possible to identify C' in a roundabout manner, with the help of special effects which occur in the presence of C', but not in the presence of C, and which are postulated by an alternative theory T'. An example of such an effect would be that C' triggers a macroprocess M (Worral has difficulties with 'triggers': any dictionary will tell him what the word means). In this case T' gives us evidence against T that could not have been discovered by using T and the associated experiments only: for God, M or C' are evidence against T; we humans, however, need T' to ascertain that fact.

Brownian motion is a special case of the situation I have just described: C are the processes in an undisturbed medium in thermal equilibrium according to the phenomenological theory of thermodynamics; C' are the processes in such a medium according to the kinetic theory. C and C' cannot be experimentally distinguished because any instrument for the measurement of heat contains the very same fluctuations it is supposed to reveal in our special case. M is the motion of the Brownian particle, T' the kinetic theory. As in the Galilean case, we can press these elements into the Duhemian scheme by saying that one auxiliary hypothesis was replaced by another auxiliary hypothesis and that some difficulty was removed thereby. Note, however, that in our case it was not the difficulty that led to the replacement but the replacement that helped us find the difficulty – and *this* feature disappears in Worral's analysis.

Turning to more general objections, I wholeheartedly agree with Ian Hacking that the sciences are more complex and many-sided than I assumed in some of my earlier writings and also in parts of AM. I had simplistic ideas both about the elements of science and about their relations. Science does contain theories – but theories are neither its only ingredients nor can they be adequately analysed in terms of statements or other logical entities. We may admit that there exist axiomatic formulations and that some scientific ideas have been defined in a precise way; we may also admit that scientists when doing research occasionally rely on the results of these efforts. However, they use them in a rather loose way, combining axioms from different domains in a manner liable to give a heart attack to philosophers entranced by simple forms of logic. Logic itself has now entered a stage where formalizations are used in a freewheeling way and where 'anthropological' considerations (finitism) play an important role. Altogether the scientific enterprise seems to be much closer to the arts than older logicians and philosophers of science (myself among them) once thought (for this side of the matter cf. my essay *Wissenschaft als Kunst*, Frankfurt 1984).

My first doubts about the identification of science with the explicit features of its theories and its observational reports arose in 1950 when I read a manuscript copy of Wittgenstein's *Philosophical Investigations*. I still expressed these doubts abstractly, in terms of conceptual problems (incommensurability; 'subjective' elements of the theory of explanation). Starting work on chapter 17 of AM, I was then led to question the adequacy of abstract procedures both in the sciences and in the philosophy of science. Here I learned from three books: Bruno Snell's magnificent *Discovery of the Mind*, recommended to me by Barbara Feyerabend; Heinrich Schaefer's *Principles of Egyptian Art*, a book of importance far beyond the subject matter treated; and Vasco Ronchi's *Optics, the Science of Vision*. Today I would add Panofsky's writings on the history of the arts (especially his pathbreaking essay *Die Perspektive als Symbolische Form*) and Alois Riegl's *Spätrömische Kunstindustrie* where the doctrine of artistic relativism is explained simply, and with powerful arguments. All I had to do to extend these arguments to the sciences was to realise that scientists, too, produce works of art — the difference being that their material is thought, not paint, nor marble, nor metal, nor melodious sound.

Regarding thought itself, I started my escape from positivism by distinguishing between two different kinds of traditions which I called abstract traditions and historical traditions respectively (details in ch. 1, Vol. 2 of my *Philosophical Papers*, and in *Wissenschaft als Kunst*, as well as in chapter 3 of the present book). There are many ways of characterising these traditions. One difference which I found a most helpful starting point is the way in which the two traditions deal with their objects (people, ideas, gods, matter, the universe, societies – and so on).

Abstract traditions formulate statements. The statements are subjected to certain rules (rules of logic; rules of testing; rules of argument – and so on) and events affect the statements only in accordance with the rules. This, it is said, guarantees the 'objectivity' of the information conveyed by the statements, or of the 'knowledge' they contain. It is possible to understand, criticize, or improve the statements without having met a single one of the objects described (examples: elementary particle physics; behavioural psychology; molecular biology which can be run by people who never in their life saw a dog, or a prostitute).

Members of *historical traditions* also use statements, but they talk in a very different way. The assume, as it were, that the objects already have a language of their own and they try to learn this language. They try to learn it not on the basis of linguistic theories but by immersion, just as small children familiarise themselves with the world. And they try to learn the language of the objects as they are, and not as they appear after they have been subjected to standardizing procedures (experiments, mathematization). Categories of the abstract approach such as the concept of an objective truth cannot describe a process of this kind which depends on the idiosyncracies of both objects and observers (it makes no sense to speak of the 'objective existence' of a smile which, depending on context, can be seen as a kind smile, a cruel smile, or a bored smile).

Abstract and historical traditions have fought each other from the beginning of Western thought. Their contest started with the 'ancient battle between philosophy and poetry' (Plato, *Republic* 607b – see chapter 3 of the present book). It continued in medicine where the theoretical approach of Empedocles and the element-physicians was criticized by the author of *Ancient Medicine* (details in chapter 1, section 6 and in chapter 6, section 1). The antagonism characterized Thucydides's criticism of Herodotus and it has survived until today — in pyschology (behaviourism versus *'verstehende'* methods), biology (molecular biology versus qualitative types of biological research), medicine ('scientific' medicine versus healers of all sorts), ecology, and even mathematics (Cantorianism versus constructivism – to use terms first suggested by Poincaré). Abstract traditions change into historical traditions during times of crisis and revolution, which supports my thesis that *good sciences are arts, or humanities, not sciences in the textbook sense*. Ian Hacking's analysis of experimental procedures is an excellent illustration of the art-aspect of scientific research.

Alan Musgrave shows that the instrumentalistic tradition in ancient astronomy was much weaker than Duhem thought. He forgot to mention that modern scientific realism uses an instrumentalism of qualitites and qualitative laws: realists take it for granted that qualities which do not enter the body of science but which enable us to contribute to it will not lead us astray. Modern science, which created but never solved the mind-body problem, uses instrumentalism at its very basis – and it shows (for instance, in the quantum theory of measurement). In a short introduction, which has nothing to do with the bulk of his paper and which he seems to have added as an afterthought, Musgrave produces a curious criticism of an earlier essay of mine (reprinted in *Philosophical Papers* Vol. 1, chapter 11). There I argued that most philosophical reasons for realism are too weak to overcome the physical reasons against it, and that they must be made stronger; and I then developed the needed stronger reasons. According to Musgrave I did the very opposite – I tried to find universal arguments for *instrumentalism*! I don't think that Alan misread what I had written, for he is a careful critic and the paper he criticizes one of the clearest papers I have written – but I am quite ready to accept a plea of temporary insanity. Let me add, incidentally, that I no longer believe in the relevance, for our understanding of the sciences, of general arguments such as those produced in my paper.

I agree with practically all the points and objections made in Grover Maxwell's beautiful essay on the mind-body problem. I admit that despite good intentions I 'too often relapsed into the empiricist . . . practice of treating . . . meaning in an apriori manner' (but I had my sane moments, too, and then I treated meanings as neurophysiological structures or 'programmes': see *Philosophical Papers* vol. 1, chapter 6; vol. 2, chapter 9). I also admit that I occasionally forgot the contingent nature of the pragmatic theory of observation (for my sane moments on this point cf. my little note 'Science without Experience', *Philosophical Papers*, vol. 1, chapter 7, which made Ayn Rand curse me in an open letter to all American philosophers). It is true that in criticizing acquaintance I 'set up a straw man'. Actually, the straw man (straw woman?) was set up not by me but by the sense-datists – but having removed her (it? him?) I thought I had removed all aspects of acquaintance – and in this I was certainly mistaken. I was not consistent in my mistake for I occasionally assumed, as Russell had done, that the brain could be directly perceived, but I did not draw the correct conclusion and declare some physical events to be mental. I am not too upset by the fact that some of my arguments may provide amunition for the eliminative mentalist – this, I think, applies to all arguments about contingent matters. Grover's own theory, on the other hand, seems to me to rely too heavily on scientific notions and procedures. Grover's assertion that 'science works' does not remove my uneasiness. Science works sometimes, it often fails and many success stories are rumours, not facts. Besides, the efficiency of science is determined by criteria that belong to the scientific tradition and thus cannot be regarded as objective judges. (For example, science does not save souls.) I conclude that Grover has shown how our notions of mind and body can be developed *within the scientific framework* without thereby removing ideas that come from different traditions (the tradition of the Dogon, or the Azande, or of Ecuadorian peasants). And I am very glad that he has not succeeded in doing the latter; at least there is now a chance of meeting him again, on a different plane, in different circumstances but, hopefully, with his caustic homour unchanged.

4 Science – one Tradition among Many

The second topic of my writings is the authority of the sciences. I assert that there exist no 'objective' reasons for preferring science and Western rationalism to other traditions. Indeed, it is difficult to imagine what such reasons might be. Are they reasons that would convince a person, or the members of a culture, no matter what their customs, their beliefs or their social situation? Then what we know about cultures shows us that there are no 'objective' reasons in this sense. Are they reasons which convince a person who has been properly prepared? Then all cultures have 'objective' reasons in their favour. Are they reasons referring to results whose importance can be seen at a glance? Then again all cultures have at least some 'objective' reasons in their favour. Are they reasons which do not depend on 'subjective' elements such as commitment or personal preference? Then 'objective' reasons simply do not exist (the choice of objectivity as a measure is itself a personal and/or group choice – or else people simply accept it without much thought).

It is true that Western science has now infected the whole world like a contagious disease and that many people take its (intellectual and material) products for granted – but the question is: was this a result of argument (in the sense of the defenders of Western science), i.e. was every step of the advance covered by reasons that are in agreement with the principles of Western rationalism? Did the infection improve the lives of those who were touched by it? The answer is no to both questions. Western civilization was either imposed by force, not because of arguments showing its intrinsic truthfulness, or accepted because it produced better weapons (see chapter 1, section 9); and its advance, while doing some good, also caused enormous damage (for a survey consult J.H. Bodley, *Victims of Progress*, Menlo Park, California 1982). It not only destroyed spiritual values which gave meaning to human lives, it also damaged a corresponding mastery of the material surroundings without replacing it by methods of comparable efficiency. 'Primitive' tribes knew how to deal with natural disasters such as plagues, floods, droughts – they had an 'immune system' that enabled them to overcome a great variety of threats to the social organism. In normal periods they exploited their environment without damaging it, using knowledge of the properties of plants, animals, climatic changes and ecological interactions which we are only slowly recovering (details and ample literature are given in Lévi-Strauss, *The Savage Mind*, and later more detailed studies of a similar kind). This knowledge was severely damaged and partly destroyed, first by the gangsters of colonialism and then by the humanitarians of developmental aid. The resulting helplessness of large parts of the so-called Third World is the result of, not a reason for, outside interference.

Majid Rahnema, an Iranian scholar, has compared the effects of developmental aid with the effects of the illness Aids which destroys the immune system of the human body (*From 'Aid' to 'Aids'*, unpublished manuscript, Stanford 1984). He has also commented on the way in which knowledge was turned from a common good to a rare and inaccessible commodity. 'Cultures and Civilizations,' he writes (*Education for Exclusion or Participation?*, manuscript, Stanford, 16 April 1985),

> were formed, enriched and transmitted by millions of people who were *learning by living and doing*, for whom living and learning was synonymous, as they had to learn for living and they learned whatever was meaningful to them and to the community they belonged to. Before the present school system came into being, for thousands of years, education was not a scarce commodity. It was not a product of some institutional factories, the possession of which could bestow upon a person the right to be called 'educated . . . The [new] school system . . . serve[d] as a rather efficient channel of sieving out, into the Power Establishment, the most ambitious – and sometimes the brightest – aiming at personal and professional fame. It also, paradoxically, did serve as a 'culture medium' to some outstanding individuals, among them radical thinkers and revolutionaries who used some of its unique learning resources for their own liberating purposes. Yet, on the whole, it soon became an 'infernal machine' which distinguished itself in the systematic organization of excluding processes against the poorest and the powerless . . . The old days . . . when 'every adult was a teacher' were over. Now, only those certified by the school system, according to its self devised criteria, could have the right to teach. *Education thus became a scarcity* [my emphasis].

It is interesting to see how little influence these discoveries have had on the sermons preached by professional rationalists. Karl Popper, for example, bemoans the 'general anti-rationalist atmosphere . . . of our time', praises Newton and Einstein as great benefactors of humanity but breathes not a single word about the crimes committed in the name of Reason and Civilization. On the contrary, he seems to think that the benefits of civilization may occasionally have to be imposed, on unwilling victims, by a 'form of imperialism' (see chapter 6, section 1).

There are various reasons why so many intellectuals still argue in this short-sighted way. One reasons is ignorance. Most intellectuals have not the foggiest idea about the positive achievements of life outside Western civilization. What we had (and, unfortunately, still have) in this area are rumours about the excellence of science and the dismal quality of everything else. Another reason lies in the immunizing moves rationalists have devised to overcome difficulties. For example, they distinguish between basic science and its applications: if any destroying was done, then this was the work of the appliers, not of the good and innocent theoreticians. But the theoreticians are not that innocent. *They* are recommending analysis over and above understanding, and this even in domains dealing with human beings; *they* extol the 'rationality' and 'objectivity' of science without realising that a procedure whose main aim it is to get rid of all human elements is bound to lead to inhuman actions. Or they distinguish between the good which science can do 'in principle' and the bad things it actually does. That can hardly give us comfort. All religions are good 'in principle' – but unfortunately this abstract Good has only rarely prevented their practitioners from behaving like bastards.

Thoughtless people are in the habit of pointing out that every 'reasonable' person will be persuaded that science knows best. The comment admits a weakness of argumentation: arguments do not work on everyone, they work only on people who have been properly prepared. And this is a general feature of all idological debates: arguments in favour of a certain world view depend on assumptions which are accepted in some cultures, rejected in others, but which because of the ignorance of their defenders are thought to have universal validity. Kekes's attempt to overcome relativism is an excellent example of this situation.

He makes three assumptions: (1) it is important to solve problems; (2) there exist more or less unambiguous methods for solving problems; (3) some problems are independent of all traditions – problems of this kind Kekes calls problems of life. Kekes also assumes that explicit conceptualisation plays an important part in recognizing, formulating, and solving problems. But for the Orphics, some Christians, and some Moslem fundamentalists, many of the things a Western intellectual might call problems were not undesirable situations waiting to be removed by human ingenuity but either tests of moral fibre (cf. the function of initiation rites), or preparations for a difficult task, or necessary ingredients of a life that would cease to be human without them. Some cultures treat problems as quirks which cause amusement, not consternation; one simply lets them pass instead of trying to 'solve' them.

White government officials in Central Africa were often upset by the fact that problems they noticed and conveyed to their black colleagues were not dealt with seriously, by an increased effort of thought, but were simply laughed out of court: the bigger the problem, the greater the hilarity. This, the white rationalists said, was a very irrational way to behave – and so it was, according to their standards. On the other hand – what a fine way of avoiding wars and the misery they create! 'Do something' certainly is not uniformly superior to 'Let it be'. Kekes articulates the procedures customary within certain traditions – he does not give us 'objective', i.e. transtraditional principles.

'Problems of life' in Kekes's sense are parts of special and relatively young traditions of a materialistic-humanistic bent. Their solutions cannot be impartial judges of the rest. Moreover, even secular solutions allow for many ways of living outside the sciences, as is shown by our artists and by the wide spectrum covered by apparently 'objective' concepts such as the concept of health (cf. Foucault). We have to admit that many values and many cultures have ceased to exist; they were killed and hardly anybody now remembers them. But this does not mean that we cannot learn from them and, besides, Kekes wants a *theoretical* solution to the problem of relativism – and such a solution is not forthcoming.

Similar remarks apply to Noretta Körtge's interesting and provocative essay. She must be praised for emphasizing that in dealing with citizens appearance is at least as important as 'reality' (which at any rate is nothing but the way in which things appear to fashionable experts): 'not only must justice be done but justice must *seem* to be done'. Well said! *What counts in a democracy is the experience of the citizens, i.e. their subjectivity and not what small gangs of autistic intellectuals declare to be real* (if an expert does not like the ideas of the common folk then all he has to do is to talk to them and try to persuade them to think along different lines; in so doing, he must not forget that he is a beggar and not a 'teacher' trying to pound some truth into the heads of penitent pupils). But her attempt to separate this experience from some 'reality' cannot succeed. I agree that the sciences, and civilizations built around them, contain something called 'expert opinion' and that it differs from what experts call 'popular superstitions' – but I would add that this is true of other traditions as well (for example, it is true of the Dogon as Griaule has shown in his marvellous book). I also agree that expert opinion occasionally shows some uniformity – all churches have temporary uniformities – but the occasional convergences in some areas are more than compensated by disagreements in others. Nor does the convergence of expert opinion establish an objective authority and if it does, then we have many different authorities to choose from: the distinction between expert-reality and layman-appearance dissolves into what appears to every one of us, experts included.

That rationalists clamouring for objectivity and rationality are just trying to sell a tribal creed of their own becomes very clear from the reactions of some less gifted members of the tribe. Thus Tibor Macham, writing at the expense of an outfit ominously called the Reason Foundation (I am referring to a review of SFS which appeared in *Philosophy of the Social Sciences*, 1982), distinguishes between acceptable standards, ideas and traditions, and traditions that are 'mere caprice and destructive to human life'. What is the rationale for his distinction? A theory of man. What is the gist of his theory of man? That 'human beings are rational animals . . . biological beings with the distinctive need and capacity for principled (or conceptual) thought and action.' This, of course, is a perfect description of intellectuals (the only thing missing is the craving for large salaries) – but a person with a somewhat different perspective will have to point out, in all modesty, that Macham's 'theory of man' is but one view among many and that intellectuals, fortunately, are still only a small percentage of humanity. There is the view that humans are misfits in the material world, unable to understand their position and their purpose and 'with a distinctive need' for salvation; there is the view, closely related to the one just mentioned, that humans consist of a divine spark enclosed in an earthen vessel, a 'trace of gold embedded in dirt' as the Gnostics were in the habit of saying, 'with the distinctive need' for liberation by faith. And these are not just abstract and 'capricious' views – they have been, and still are, part of the lives of millions of people. There is the view, found among Buddhists, that humans want to escape pain, that thought and purposeful action based on thought are the main causes of pain, and that pain will cease once customary distinctions are removed and customary purposes abolished. The Hopi Genesis represents humans as being originally in harmony with Nature. Thought and striving, or in other words the very same 'need for principled thought and action' which Macham makes the centre of humanity, destroy the original harmony, the animals withdraw from the humans, the human species is split into races, tribes and small groups with different ideas, and different languages arise until even individuals no longer understand each other. But humans, 'having the distinctive need and capacity for' harmony, can overcome the fragmentation by freeing themselves from the fetters of conceptual thought and the strife it creates and by basing their lives on love and intuitive understanding.

There are numerous views of this kind and they all differ from the theory mentioned *and taken for granted* by Macham. Now it is of course Macham's good right to favour one view and condemn another. But he does so posing as a rationalist and a humanitarian. He claims to have not only anathemas but also arguments and to be motivated by a love for humanity. A look at his criticism shows that both claims are suprious. His arguments are but curses pronounced in the stiff rhetoric of the self-conscious scholar and his love for humanity stops right at his office door (or at the cashier's desk of the Reason Foundation).

As is customary among intellectuals, Macham uses unanalysed cases such as the Jonestown killings to frighten his readers instead of trying to enlighten them (German 'rationalists' use Auschwitz and, more recently, terrorism *ad nauseam* for the same purpose). 'These are easy cases', says Macham. How naive can you get? In Jonestown, some people committed suicide freely, in the full knowledge of what they were doing (case 1). Others wavered, were undecided, would have liked to survive but submitted to the pressure of their peers and their leaders (case 2). Still others were simply murdered (case 3). For Macham the distinctions do not exist. But they are essential for an instructive analysis of the case. Case 3 may be 'easy' if one wants to talk in this superficial way, though there are sizeable problems even here (should one kill bodies to save souls? Rational Inquisitors thought so and with excellent arguments: are these arguments to be disregarded? Are we to take materialism for granted? I have no objection to the latter step – but where does this leave a rationalist, i.e. a person who claims to have arguments for every move he makes?). Case 1 is again 'easy' though not in the way assumed by Macham. Of course, it is 'destructive to human life' – but is human life an overriding value? The Christian martyrs did not think so and neither Macham nor any other rationalist has succeeded in showing that they were mistaken. They had a different opinion – that is all. Socrates expressed a similar sentiment before dying; he was not alone, for the same sentiment can be found in Herodotus, in Sophocles and in other outstanding representatives of classical Greece. Not once does it occur to Macham that his view of a human being is one among many, that he is a party to the debate and not its supervisor.

There remains case 2; here I fully agree with those who demand that people be protected from peer group and leader pressures. But this caveat applies not only to religious leaders such as the Reverend Jones *but also* to secular leaders such as philosophers, Nobel Prize Winners, Marxists, Liberals, hitmen of foundations and their educational representatives: the young must be strengthened against being imposed upon by so-called teachers, and especially against ratiofascists like Macham and his peers. Unfortunately contemporary education is far from agreeing with this principle.

Finally there is the old argument that non-scientific traditions have already had their chance, that they did not survive the confrontation with science and rationalism and that attempts to revive them are therefore both irrational and unnecessary. Here the obvious question is: were they eliminated on rational grounds, by letting them compete with science in an impartial and controlled way, or was their disappearance the result of military (political, economic etc.) pressures? And the reply is almost always: the latter. The American Indians were not asked to present their views, they were first christianised, then sold out of their land and finally herded into reservations amidst a growing scientific-technological culture. Indian medicine (which was commonly used by medical practitioners of the nineteenth century) was not tested against the new pharmaceuticals that invaded the market, it was simply forbidden as belonging to an antediluvean age of healing. And so on.

Reference to past opportunities also overlooks the point that even clear and unambiguous refutations do not seal the fate of an interesting point of view (for what follows, cf. SFS, pp. 100ff., and chapter 1, section 1 of the present book); the means of refutation (experimental equipment, the theories used for the interpretation of the results obtained) constantly change, and with them the nature of the argument. One should also note the striking similarity between the argument from success and comments such as those made by the Nazis after their triumph in 1933: liberalism had already had its chance, it was defeated by the national forces and it would be silly to try to reintroduce it.

Finally, it is up to the citizens to choose the traditions they prefer. Thus democracy, the fatal incompleteness of criticism, and the discovery that the prevalence of a view never is and never was the result of an exclusive application of rational principles, all suggest that attempts to revive old traditions and to introduce anti-scientific views are to be praised as the beginnings of a new age of enlightenment, where our actions are guided by insight and not merely by pious and often quite moronic slogans.

5 Reason and Practice

What I have said so far can be summarized in the following two statements:

(A) the way in which scientific problems are attacked and solved depends on the circumstances in which they arise, the (formal, experimental, ideological) means available at the time and the wishes of those dealing with them. There are no lasting boundary conditions of scientific research.

(B) the way in which problems of society and the interaction of cultures are attacked and solved also depends on the circumstances in which they arise, the means available at the time and the wishes of those dealing with them. There are no lasting boundary conditions of human action.

Thus I criticized the view which I shall call (C), that science and humanity must conform to conditions that can be determined independently of personal wishes and cultural circumstances. And I objected to the assumption, (D), that it is possible to solve problems from afar, without participating in the activities of the people concerned.

(C) and (D) are the core of what one might call *the intellectualistic approach to (science and) social problems*. They are a matter of course for academic Marxists, liberals, social scientists, businessmen, politicians eager to help 'underdeveloped nations' and prophets of 'new ages'. Every writer who wants to improve knowledge and save humanity and who is dissatisfied with existing ideas (reductionism, for example) thinks that salvation can only come from a new *theory* and that all that is needed to develop such a theory are the right books and a few clever ideas.

(C) and (D) have also been used to discredit what I say about politics. According to my critics, I make a lot of noise but achieve little. My approach, they say, is entirely negative. I object to certain procedures – but I have nothing to offer in their stead. Marxists have been especially incensed at my mocking disregard for their two favourite playthings, Western science and humanitarianism.

These remarks are certainly correct. I have indeed no positive suggestions to make. But the reason is not that I have forgotten about the matter, or cannot compete with the speculative talents of my fellow academics – the reason is my respect for the traditions I am supposed to bless with my intellectual gifts. These traditions are historical traditions, not abstract traditions (see above, sections 2, 3 and 4, and chapter 3). Historical traditions cannot be understood from afar. Their assumptions, their possibilities, the (often unconscious) wishes of their bearers can be found only by immersion, i.e. *one must live the life one wants to change*. Neither (C) nor (D) apply to historical traditions. Boundary conditions and solutions invented by distant speculators can still be imposed *but only by disregarding the full humanity of the victims*. Intellectuals who support the imposition are not unaware of the 'human dimension'; they have 'theories of man' and they use them as guides to their actions. But these theories do not reflect their victims; they reflect the mentality of the place where they arose – university offices and seminar rooms, mainly (cf. my remarks on Tibor Macham in section 4 above): my main objection against intellectual solutions of social problems is that they start from a narrow cultural background, ascribe universal validity to it and use power to impose it on others. Is it surprising that I want to have nothing to do with such ratiofascistic dreams? Helping people does not mean kicking them around until they end up in somebody else's paradise, helping people means trying to introduce change *as a friend*, as a person, that is, who can identify with their wisdom *as well as* with their follies and who is sufficiently mature to let the latter prevail: *an abstract discussion of the lives of people I do not know and with whose situation I am not familiar is not only a waste of time, it is also inhumane and impertinent.*

It is a waste of time because the practical application of the theories found will always have to be preceded by numerous changes which may wipe out the basic programme. It is impertinent: being unfamiliar with the conditions of the strangers, with the ways in which these conditions appear to them, having no direct experience of their dreams, fears, desires, I refuse to make my own standards, my so-called knowledge (whether puny or impressive – that does not matter), my own very limited humanity the basis of 'objective' diagnoses and suggestions (only very naive or intolerant people can believe that a study of the 'nature of man' is superior to personal contacts, in one's private lives as well as in politics). Jutta, who bears a woman's name but who seems bent on outdoing the chauvinism of the most pushy of her male academic colleagues, says that I am lacking in heart and imagination. On the contrary: *I* can imagine that there are situations of which I have never thought, which are not described in books, which scientists have never encountered and would not recognize when confronted with. I believe that such situations occur quite frequently; I can also imagine that such situations look different to different people, affect them in different ways, raise hopes, fears, emotions I have never felt, and I have the heart to subject my distant guesses to the impressions of those immediately concerned. Jutta says I should 'examine' with 'respect' what I do not know. Examine? If I love a woman and want to share her life, for my benefit and perhaps also for hers, then I shall not 'examine' that life, whether respectfully or with disdain, I shall try to *participate* in it (provided she lets me) so that I can understand it from within. Participating in her life, I change into a new person with new ideas, feelings, ways of seeing the world. Of course, I shall make lots of suggestions – I may even drive her nuts with all my talk *but only after the change has occurred* and on the basis of the new *and shared* sensibilities it has created. Now politics, as I understand it, is in many ways related to love. It respects people, considers their personal wishes, does not 'study' them whether by polls or by anthropological field work but again tries to understand them from within, and connects suggestions for change with the thoughts and the emotions that flow from such an understanding. In a word: *politics, rightly understood, is firmly 'subjective'*. It is impossible to develop 'objective' theoretical schemes for it.

6 Elements of a Free Society

How is this account related to my ideas about the police, the equality of traditions, the separation of state and science? The answer has already been given in SFS and EFM (EFM, p. 77 and passim): ideas such as these must pass the filter of the traditions (of the citizens' initiatives) for which they were developed. A fundamental error of almost all the papers that deal with this part of my writings — and that includes the paper by Christiane van Briessen, who got my number in many other respects — is that they interpret my suggestions as if they should be read in the same way as politicians, philosophers, social critics, and 'great' men and women of all sorts want to be read: they interpret them as the outlines of a new social order which must now be imposed on people with the help of education, moral blackmail, a nice little revolution and treacly slogans (such as 'The Truth will Make you Free'), or by utilizing the pressures issuing from already existing institutions. But dreams of power such as these are not only very far from my mind – they positively make me sick. I have little love for the educator or moral reformer who treats his wretched effusions as if they were a new sun brightening the lives of those living in darkness; I despise so-called teachers who try to whet the appetite of their pupils until, losing all self-respect and self-control, they wallow in truth like pigs in the mud; I have only contempt for all the fine plans to enslave people in the name of 'god' or 'truth' or 'justice' or other abstractions, especially as their perpetrators are too cowardly to accept responsibility for these ideas but hide behind their alleged 'objectivity'. Many of my readers seem to regard such machinations as a very normal procedure – how else can I explain that they read my proposals in this manner? But the loose and sketchy remarks on the state, on ethics, on education, and on the business of science which I made in AM and SFS must be examined by the people for whom they are meant. They are subjective opinions, not objective guidelines; they are to be tested by other subjects, not by 'objective' criteria, and they receive political power only after everybody concerned has considered them: the consensus of those addressed, not my arguments, finally decides the matter.

The objection that people must first be taught to think only reflects the conceit and the ignorance of its authors, for the basic problem is: who can talk and who should remain silent? Who has knowledge and who is merely obstinate? Can we trust our experts, our physicists, our philosophers, our healers, our educators, do they know what they are talking about or do they merely want to duplicate their own miserable existence? Have our great minds, have Plato, Luther, Rousseau, Marx, anything to offer, or is the reverence we feel for them merely a reflection of our own immaturity?

These questions concern all of us – and all of us must participate in their solution. The most stupid student and the most cunning peasant; the much honoured public servant and his long-suffering wife; academics and dog catchers, murderers and saints – they all have the right to say: look here, I, too, am human; I, too, have ideas, dreams, feelings, desires; I, too, have been created in god's image – but you never paid attention to my world in your pretty tales (it was different in the Middle Ages; cf. Friedrich Heer, *Die Dritte Kraft*, Frankfurt 1959). The relevance of abstract questions, the content of the answers given, the quality of life adumbrated in these answers – all these things can be decided only if everyone is permitted to participate in the debate and encouraged to give her or his views on the matter. The best and simplest outline of the ideas just explained is found in Protagoras's great speech (Plato, *Protagoras*, 320c-328d): the citizens of Athens do not need any instruction in their language, in the practice of justice, in the treatment of experts (warlords, architects, navigators); having grown up in an open society where learning is direct and not mediated and disturbed by educators, they learned all these things from scratch. As for the further objection that states and citizens' intitiatives do not arise out of the blue but must be set in motion by purposeful action — that is easy to answer: let the objector start a citizens' initiative, and he will soon find what he needs, what furthers his ambitions, what obstructs them, to what extent his ideas are a help to others, to what extent they hinder them, and so on.

This, then, is my answer to the various criticisms of 'my' 'political model'. The model is vague – very true – but the vagueness is necessary, for it is supposed to 'make room' (EFM, p. 160) for the concrete decisions of those using it. The model recommends an equality of traditions: any proposal must first be checked by the people for whom it is meant; nobody can foresee the result. (The pygmies, for example, or the Mindoro of the Philippines, do not want equal rights – they just want to be left alone.) Conflicts are not dealt with by 'education' but by a police force. Margherita von Brentano interprets the last suggestion as implying that citizens may only talk and perhaps write but that their actions are severely restricted. Other critics have thrown up their hands in despair: speak of the police – and liberals and Marxists alike are liable to wet their pants. This is precisely the mistake described above. For the police is not an external agent that pushes the citizens around; it is introduced by citizens, consists of citizens and serves their needs (cf. my comments on the protective guards of the Black Muslims, EFM, p. 162, p. 297). Citizens do not just think, they decide about everything in their surroundings. I merely suggest that it is more humane to regulate behaviour by external restrictions – such restrictions can be easily removed when found impractical – than to improve souls. For assume we succeed in implanting The Good in everybody – how then shall we ever be able to return to Evil?

7 Good and Evil

With this remark I come to a point which has enraged many readers and disappointed many friends – my refusal to condemn even an extreme fascism and my suggestion that it should be allowed to survive. Now one thing should have been clear: fascism is not my cup of tea (cf. EFM, 156: 'despite my own very widely developed sentimentality and my almost instinctive tendency to "act in a humanitarian manner" '). *That* is not the problem. The problem is the *relevance* of my attitude: is it an inclination which I follow and welcome in others; or has it an 'objective core' that would enable me to combat fascism not just because *it does not please me*, but because *it is inherently evil*? And my answer is: we have an inclination – nothing more. The inclination, like every other inclination, is surrounded by lots of hot air and entire philosophical systems have been built on it. Some of these systems speak of objective qualities and of objective duties to maintain them. But my question is not how we speak but what content can be given to our verbiage. And all I can find when trying to identify some content are different systems asserting different sets of values with nothing but our inclination to decide between them (SFS, part 1). Now if inclination opposes inclination then in the end the stronger inclination wins, which means, today, and in the West: the bigger banks, the fatter books, the more determined educators, the bigger guns. Right now, and again in the West, bigness seems to favour a scientifically distorted and belligerent (nuclear weapons!) humanitarianism – and so the matter has come to a temporary rest at this point.

This, incidentally, was one of the lessons I learned from the life of Remigius, the inquisitor. Margherita von Brentano, who mentions my reference to him, was kind enough not to assume that I am pleading for a revival of witchcraft and witchcraft persecutions. Of course, this is not my intention. Nor do I think I would remain a silent witness of such persecutions. But my explanation would be that the matter does not please me and not that it is inherently evil or based on a backward view of the universe. Such expressions far exceed what can be supported by the best intentions and the most clever arguments. They give the user an authority he simply does not possess. They put him on the side of the angels when all he does is to express his personal opinions. Truth herself seems to be his companion when again we are dealing only with an opinion and a very badly argued one at that. There existed lots of arguments against atoms, the motion of the earth, the aether – and yet all these things returned to the scene. The existence of God, the Devil, heaven, hell was never attacked with even half-way decent reasons. Thus if I want to remove Remigius and the spirit of his times then I can of course proceed to do so, but I must admit that the only instruments available to me are the powers of rhetoric and self-righteousness. If, on the other hand, I accept only 'objective' reasons, then the situation forces me to be tolerant, for there are no such reasons, in this case any more than in others (SFS, parts 1 and 2; EFM, chapter 3).

Remigius believed in God, he believed in an afterlife, in hell and its tortures and he also believed that the children of witches who were not burned would end up in hell. He did not just believe those things, he could have provided arguments. He would not have argued in our manner and his evidence (the Bible, the Church Fathers, the decisions of Church councils etc.) would not have been what we call evidence. But this does not mean that his ideas were without substance. For what do *we* have to oppose him? The belief that there is a scientific method and that science is successful? The first part of the belief is false (cf. section 2 above); the second part is correct but must be supplemented by saying that there were and still are many failures as well and that the successes occur in a narrow domain that hardly touches what is at issue here (the soul, for example, never enters the scene). What falls outside the domain, such as the idea of hell, was never *examined*, it was *lost*, just as the scientific achievements of antiquity were lost by the early Christians.

Within the framework of his thought Remigius acted as a responsible and rational human being and he should be praised, at least by rationalists. If we are repelled by his views and unable to give him his due then we must realise that there are absolutely no 'objective' arguments to support our repulsion. We can of course sing moral arias, we may even write an entire opera where these arias hang together beautifully – but we cannot build a bridge from all that noise to Remigius and, appealing to *his* reason, bring him over to our side. For he does use his reason, but with a different purpose, according to different rules and on the basis of different evidence. There is no way out: *we bear full responsibility for not proceeding as Remigius does and no objective values will plead our case should we discover that our actions have led to disaster*.

On the other hand, let us not forget our own inquisitors, our scientists, physicians, educators, sociologists, politicians, 'developers'. Just look at those physicians who until quite recently cut, poisoned and irradiated without having examined alternative methods of treatment which were well known, had no dangerous consequences and could claim to be successful. Was it not worth trying such methods (was it not worth trying to keep the children of witches alive)? It was worth trying. But all we heard in reply was: *anathema sit*! Or let us examine the efforts of our educators who year in year out are let loose on the younger generation to fill it with 'knowledge' without regard for the background of the pupils. Entire cultures have been killed, their immune systems destroyed (cf. section 4), their knowledge turned into a scarcity – and all that in the name of progress (and money, of course): the spirit of Remigius, my dear Margherita von Brentano, is still with us, in economics, in energy production and (mis)use, in foreign aid, in education, the important difference being that Remigius acted for humanitarian reasons (he wanted to save little children from eternal damnation) while his modern successors only care for their professional integrity: they not only lack perspective, they also lack humanity. I don't like them, either — but here my motives are again not objective standards, but dreams of a better life.

Now if one combines such dreams (which I have) with an idea of objective values (which I reject) and calls the result a moral conscience then *I have no moral conscience* and fortunately so, I would say, for most of the misery in our world, wars, destruction of minds and bodies, endless butcheries are caused not by evil individuals but by people who have objectivised their personal wishes and inclinations and thus have made them inhuman.

This, incidentally, is the only thing Agassi seems to have noticed in his strange outburst. Agassi says he will speak the truth. That is nice of him but does not give us much comfort. For as critics of his scientific work have pointed out, long ago, he only rarely knows what he is talking about even when he is trying to tell the truth (for example: item 882 in Rosen's Copernicus bibliography, *Three Copernican Treatises*, New York 1971). His paper confirms the impression. He says I volunteered for the German army – I was drafted. He says I tried to forget the political and moral aspects of the Second World War – I did not notice them; at eighteen I was a book-worm not a *mensch*. He says I idolised Popper. Now it is quite true that I like to idolize people, I like to be able to look up to somebody, to admire her, to take her as an example – but Popper is not the stuff idols are made of. Agassi calls me a disciple of Popper. This is true in one sense, quite untrue in another. It is true that I listened to Popper's lectures, sat in his seminar, occasionally visited him and talked to his cat. This I did not of my own free will but because Popper was my supervisor: working with him was a condition of my being paid by the British Council. I had not chosen Popper for this job, I had chosen Wittgenstein and Wittgenstein had accepted. But Wittgenstein died and Popper was the next candidate on my list. Also, doesn't Agassi remember how often he begged me, on his knees, to give up my *reservatio mentalis*, fully to commit myself to Popper's 'philosophy' and, especially, to spread lots of Popper-footnotes all over my essays? I did the latter – well, I am a nice guy and quite willing to help those who seem to live only when they see their name in print – but not the first: at the end of the year Agassi is speaking of (1953), Popper asked me to become his assistant; I said no despite the fact that I had no money and had to be fed now by the one, now by the other of my more pecunious friends.

Agassi also produces some of the rumours which are apparently needed to make life in the Popperian Church bearable: he quotes Popper as saying that I once tearfully regretted having participated in the Second World War. That is quite possible, I am an emotional person and have done many stupid things in my life – but it is unlikely: I never discuss personal matters with strangers and, besides, there was nothing to be sorry about except perhaps insufficient intelligence in the attempt to escape the draft. The tears, most likely, were tears of boredom which flowed rather freely during my visits to the Master. It is a sad sign of the decay of standards of scholarship in Germany that a piece of lachrymose trash like Agassi's essay could be written with the aid of a stipend that bears the old and honourable name of Alexander von Humboldt.

There is only one point where Agassi shows some grasp of reality and this concerns our discussion of moral issues. I remember the discussion well. Agassi urged me to take a stand, i.e. to sing moral arias. I felt very uncomfortable. On the one hand the matter seemed quite idiotic – I sing my aria, the Nazi sings his – now what? On the other hand I felt the irrational pressure of Auschwitz which Agassi and many ideological street singers before and after him have used shamelessly to urge people into empty gestures (or to brainwash them so that the gestures receive 'meaning'). What do I say today?

I say that Asuchwitz is an extreme manifestation of an attitude that still thrives in our midst. It shows itself in the treatment of minorities in industrial democracies; in education, education to a humanitarian point of view included, which most of the time consists in turning wonderful young people into colourless and self-righteous copies of their teachers; it becomes manifest in the nuclear threat, the constant increase in the number and power of deadly weapons and the readiness of some so-called patriots to start a war compared with which the holocaust will shrink into insignificance. It shows itself in the killing of nature and of 'primitive' cultures with never a thought spent on those thus deprived of meaning for their lives; in the colossal conceit of our intellectuals, their belief that they know precisely what humanity needs and their relentless efforts to recreate people in their own, sorry image; in the infantile megalomania of some of our physicians who blackmail their patients with fear, mutilate them and then persecute them with large bills; in the lack of feeling of many so-called searchers for truth who systematically torture animals, study their discomfort and receive prizes for their cruelty.

As far as I am concerned there exists no difference whatsoever between the henchmen of Auschwitz and these 'benefactors of mankind' – life is misused for special purposes in both cases. The problem is the growing disregard for spiritual values and their replacement by a crude but 'scientific' materialism, occasionally even called humanism: man (i.e. humans as trained by their experts) can solve all problems – they do not need any trust in and any assistance from other agencies. How can I take a person seriously who bemoans distant crimes but praises the criminals in his own neighbourhood? And how can I decide a case from afar

seeing that reality is richer than even the most wonderful imagination.

It is one thing to be in the forefront of the fight against cruelty
and oppression, for then you can see and smell your enemy; and
your whole being, not only your ability to rhapsodize, will be
engaged in the attempt to defeat him. It is quite a different thing
to shake one's head and to decide about Good and Evil while
sitting in a comfortable office. I know – many of my friends can
make such a decision with both hands tied behind their back –
they obviously have a well developed moral conscience. I, on the
other hand, taking the distance seriously, would like to consider
a different view where Evil is part of Life just as it was part of
Creation. One does not welcome it – but one is not content with
infantile reactions either. One delimits it – but one lets it persist
in its domain. For nobody can say how much good it still contains
and to what extent the existence of even the most insignificant
good thing is tied up with the most atrocious crimes.

8   Farewell to Reason

What was the origin of the criticisms on which I have commented
in this chapter? And why did I write a reply?

It is easy to answer the first question.

About eight years ago (1979), Hans Peter Doerr was invited to
become an author of the famous Suhrkamp publishing house in
Germany. He refused, for he had other obligations. But he also
had a bad conscience – it is not easy for Hans Peter to reject
friendly invitations. Dr. Unseld, the guiding spirit of the
Suhrkamp publishing house whose ability to sniff out the bad
conscience of people is only exceeded by his expertise in mani-
pulating it, discovered Hans Peter's predicament and treated it
with words, food, and drink. Result: Hans Peter conceived the
idea of a PKF festival and started sending letters in all directions.
Some of the letters were returned unopened, others with reflections on his sanity, still others with the customary excuse of lack
of time – but quite a few people decided to praise me or to curse
me, or to exorcise me by surrounding me with rhetorical circles.
Thus it was not the merit of my 'work' that led to this collection,
but the power of alcohol.

It is much more difficult to answer the second question. Many
people — scientists, artists, lawyers, politicians, priests — draw

no distinction between their profession and their lives. If they are successful, then they take this as an affirmation of their very existence. If they fail in their profession, then they think they have failed as human beings as well, no matter how much joy they may have given to their friends, children, wives, lovers, dogs. If they write books, be they novels, collections of poems, or philosophical treatises, then these books become part of an edifice built from their very substance. 'Who am I?' Schopenhauer asked himself – and he replied, 'I am the person who wrote *The World as Will and Idea* and solved the great problem of being.' Parents, brothers, sisters, husbands, mistresses, parakeets (budgerigars for my British readers), even the most personal feelings of the author, his dreams, fears, expectations, have meaning only with respect to that edifice and they are described accordingly: the wife, well, she knew how to cook, to clean, to wash and to create the right atmosphere; the friends, well, they understood the poor chap during trying times and gave him support, they lent him money, they eagerly helped with the birth of the monsters he brought forth – and so on and so forth. This attitude is widespread. It is the basis of almost all biographies and autobiographies. It is found in really great thinkers (Socrates, a few hours before his death, gets rid of his wife and children so that he can chat about profound things with his adoring students: *Phaedo* 60a7. The artistic parallel is told with gusto and much hatred by Claire Goll in her autobiography, *Ich verzeihe keinem*, Munich 1980), but it is also quite common among the academic rodents of today.

To me this attitude is alien, incomprehensible and slightly sinister. True, I, too, once admired the phenomenon from afar; I hoped to enter the castles from whence it spread and to participate in the wars of enlightenment the learned knights had started all over the world. Eventually I noticed the more pedestrian aspects of the matter: the fact, that is, that the knights – the professors – serve masters who pay them and tell tham what to do: they are not free minds in search of harmony and happiness for all, they are civil servants (*Denkbeamte*, to use a marvellous German word) and their mania for order is not the result of a balanced inquiry, or of a closeness to humanity, it is a professional disease. So while I made full use of the sizeable salaries I got for doing very little, I was careful to protect the poor humans (and, in Berkeley, dogs, cats, racoons, even a monkey now and then) who came to my lectures from the disease. After all, I said to myself, I have some kind of responsibility for these people and

I must not misuse their trust. I told them stories and I tried to strengthen their natural contrariness, for this, I thought, would be the best defence against the ideological street singers they were about to meet: *the best education consists in immunizing people against systematic attempts at education.* But even these friendly considerations never established a closer bond between me and my job. Frequently, when driving by the university, be it now in Berkeley, or in London, or in Berlin, or here in Zürich where I am paid in solid Swiss Francs, I was startled at the thought that I was 'one of them'. 'I am a professor,' I said to myself – 'impossible – how did it happen?'

Concerning my so-called 'ideas' my attitude was exactly the same. I always liked to debate with friends, about religion, the arts, politics, sex, murder, the theatre, the quantum theory of measurement and many other topics. In such discussions I took now one, now another position: I changed positions — and even the shape of my life — partly to escape boredom, partly because I am counter suggestive (as Karl Popper once sadly remarked) and partly because of my growing conviction that even the most stupid and inhumane point of view has merit and deserves a good defence. Almost all my written . . . well, let us call it 'work', starting with my thesis, arose from such live discussions and shows the impact of the participants. Occasionally I believed I had thoughts of my own – who does not now and then become a victim of such delusions? – but I would have never dreamt of regarding these thoughts as an essential part of myself. I, so I said when considering this matter, am very different indeed from the most sublime invention I have produced and the most deeply felt conviction that pervades me, and I must never permit these inventions and convictions to get the upper hand and to turn me into their obedient servant. I might even 'take a stand' (though the practice and even the phrase with its Puritanical connotations put me off), but when I did so, then the reason was a passing whim, not a 'moral conscience' or any other nonsense of that kind.

There was another element hidden behind my unwillingness to 'take a stand', and I have discovered it only recently. I wrote AM partly to tease Lakatos (who was supposed to write a reply but died before he could do so) and partly to defend scientific practice from the rule of philosophical law. Having absorbed Ernst Mach when about fifteen and having been a student of Hans Thirring and Felix Ehrenhaft in physics, I took it for granted that the work of scientists was self-supporting and did not need any outside legitimation. I got impatient with people who though lacking any experience of the complexity of scientific research still claimed to know what it was all about and how it could be improved. I guess I was a kind of a scientific libertarian and my battle cry could have been 'leave science to the scientists!' Of course, I had once been a rationalist myself – but it needed only a simple practical example, it needed only Prof. von Weizsäcker's concrete arguments, in Hamburg, back in 1965 (I believe), to reveal the shallowness of rationalistic orations and to make me return to Mach.

There was a second experience that had a tremendous influence upon me. I repeat it in the words in which I first described it (SFS, 118f):

In the years 1964ff Mexicans, Blacks, Indians entered the university as a result of new educational policies. There they sat, partly curious, partly disdainful, partly simply confused, hoping to get an 'education'. What an opportunity for a prophet in search of a following! What an opportunity, my rationalist friends told me, to contribute to the spreading of reason and the improvement of mankind! What a marvellous opportunity for a new wave of enlightenment! I felt very differently. For it dawned on me that the intricate arguments and the wonderful stories I had so far told to my more or less sophisticated audience might be just dreams, reflections of the conceit of a small group who had succeeded in enslaving everyone else with their ideas. Who was I to tell these people what and how to think? I did not know their problems though I knew they had many. I was not familiar with their interests, their feelings, their fears though I knew they were eager to learn. Were the arid sophistications which philosophers had managed to accumulate over the ages and which liberals had surrounded with schmaltzy phrases to make them palatable the right thing to offer to people who had been robbed of their land, their culture, their dignity and who were now supposed to absorb patiently and then to repeat the anaemic ideas of the mouthpieces of the oh so human captors? They wanted to know, they wanted to learn, they wanted to understand the strange world around them – did they not deserve better nourishment? Their ancestors had developed cultures of their own, colourful languages, harmonious views of the relation between man and man and man and nature whose remnants are a living criticism of the tendencies of separation, analysis, self-centredness inherent in Western thought . . . These were the ideas that went through my head as I looked at my audience and they made me recoil in revulsion and terror from the task I was supposed to perform. For the task – this now became clear to me – was that of a very refined, very sophisticated slavedriver. And a slavedriver I did not want to be.

This experience was similar in nature to my experience vis-a-vis physics. There, too, I had strongly felt the superficiality and presumptions of a philosophy that wanted to interfere with a well formed practice. However while science is only part of culture and needs other ingredients to arrive at a full life, the traditions of my audience had been complete from the very beginning. Thus the interference was much more serious and much stronger resistance was needed. Trying to build up such resistance I considered intellectual solutions, that is, I still took it for granted that it was up to me and the likes of me to devise policies for other people. Of course, I intended to devise much better policies than those imposed by President Johnson and his aides but in doing so I, like he, took responsibility away from those I wanted to help, I dealt with them as if they were not capable of taking care of themselves. It seems that I was aware of this contradiction and it was this unconscious awareness that made me act in a distant and unconcerned way and made me refuse 'to take a stand'.

Now comes the third experience on my path – my acquaintance with Grazia Borrini, a gentle but determined fighter for peace and self reliance. Grazia had studied physics, as I had. Like me she had found this study too confining. But while I was still using abstractions (such as the idea of a 'free society') to arrive at a wider and more humane point of view, her ideas were part of 'historical traditions' (to relapse into my own constipated manner of speaking). I *did* know about these traditions and I had written about them even before I met Grazia, but again it needed a concrete encounter to make me realise what that implied. Grazia also gave me books and papers written by outstanding scholars dealing with the problems of economic and cultural (ex)change. This was a real find. First, I now had much better examples of the limits of a scientific approach than those I had been in the habit of using (astrology, voodoo, a little bit of medicine). Secondly I realised that my efforts had not been in vain and that it needed only a slight change in attitude to make them effective, both in my own eyes, and in the eyes of others. You *can* help people by writing books. I was very surprised and deeply moved when I noticed that people from different cultures whose actions I respected had read some of the things I had written and had welcomed them. So, I finally gave up my self-cynicism and decided to write one last, but good book, for Grazia, because I know her and because I write best when I have a smiling face before me (remember, I wrote AM with Imre Lakatos in mind), and, through her, for all the people who despite hunger, oppression, wars try to survive and to achieve a little bit of dignity and happiness. Of course, to write such a book I shall have to cut the remaining strings that still tie me to the abstract approach or, to revert to my usual irresponsible way of talking, I shall have to say

FAREWELL TO REASON.

Index